SOTHIS PRESS

Sothis Press aims to completely reimagine the public domain book with an emphasis on universal access to captivatingly designed print editions. We believe that technology will enable every single book ever published to be perpetually accessible in beautiful print form. Sothis is beauty in ubiquity.

www.sothis.press
1 Waterhouse Square, London, ECIN 2ST

Preparatory Studies for Deductive Methods in Storm and Weather Predictions

Clevcland Abbe

SOTHIS PRESS

ISBN 978-1-0368-0105-2

Preparatory Studies for Deductive Methods in Storm and Weather Predictions is in the public domain.

This edition is published by Sothis Press © 2023. The company is registered in England and Wales. The Company Number is 14900389.

ANNUAL REPORT

OF THE

CHIEF SIGNAL OFFICER

OF THE ARMY

TO

THE SECRETARY OF WAR

FOR

THE YEAR 1889.

IN TWO PARTS.

PART II.

WASHINGTON:
GOVERNMENT PRINTING OFFICE.
1890.

ANNUAL REPORT OF THE CHIEF SIGNAL OFFICER FOR 1889.

APPENDIX 15.

PREPARATORY STUDIES

FOR

DEDUCTIVE METHODS

IN

STORM AND WEATHER PREDICTIONS,

BY

PROFESSOR CLEVELAND ABBE, A. M., LL. D.,

Professor and Assistant in the Office of the Chief Signal Officer, Member of the National Academy of Sciences, etc.

Prepared under the direction of

BRIGADIER-GENERAL A. W. GREELY,

CHIEF SIGNAL OFFICER,

BY AUTHORITY OF THE SECRETARY OF WAR.

WASHINGTON:

GOVERNMENT PRINTING OFFICE.

1890.

CONTENTS.

	Page.
INTRODUCTION	7
CHAPTER I.—Matters that are important and unimportant	11
CHAPTER II.—Friction and vortex motions	20
CHAPTER III.—Horizontal turbulent flow of air	33
CHAPTER IV.—Ascending motions in lower strata due to buoyancy	60
CHAPTER V.—Clouds and their horizontal motions	85
CHAPTER VI.—Clouds and their buoyancy, and rain	97
CHAPTER VII.—Mechanism of storms	131

NOTE.

The text of this paper has been published during the absence of Professor Abbe and the proof read by comparison with two complete manuscripts left by the author. All illustrations not absolutely essential to the main topic were necessarily omitted, but Professor Abbe failed to make the corrections corresponding in the text. It is impossible to delay the publication. Hence, reference is made in this paper to the following figures which are not reproduced: Nos. 1, 2, 12, 13, 14, 15, 16, 17, 20, 21, 22, 25, 26, 27, 32, 33, 34, 36, 37, 38, 39, 40, 41, 42, 43, 44, 45, 46, 47, 48, 53, 54, 57, 58, 59, 60, 61, 62, 64, 65, 66, 68, 69, 70, 71, 72, 73, 74, 78, 79, 80, 81, 82.

INTRODUCTION.

Meteorology is essentially the application of hydrodynamics and thermodynamics to the atmosphere of the earth, but before we apply modern analytical methods to our problems we must, by preliminary studies, ascertain what are the conditions of the problems.

The object of the present study is to consider the physical principles that are involved in the formation and motion of storms, and that have generally guided me in making storm predictions, and without introducing mathematical analysis give a foundation for such formulæ, numerical tables or graphic methods as may eventually make an approximately complete rational system in meteorology, a deductive or philosophical method, in which the pertinent physical laws shall be followed out to their conclusions and the mechanism of a storm developed in a manner such that by future improvements the methods of meteorology shall become comparable in accuracy and perhaps in elegance to the analysis that is brought to bear on the problems of other exact sciences.

The general principle that the condensation of moisture controls the formation, growth, and internal motion of ordinary storms was early evident to many students of nature as well as to Espy and Ferrel, and has been independently worked out by John Eliot of the Meteorological Office of India; but the movement of the storm as a whole was very generally attributed to the general movement of the atmosphere; I believe that I was never willing to concede this as regards the large storms, however true it might sometimes be of the smaller thunder-storms. In 1867, in daily intercourse with Prof. Joseph Henry, I maintained the principle that a general storm moves towards the region where present conditions are favoring the greatest formation of precipitation, whether in the form of rain, cloud, or snow, and though at that time the reason why this should be so was not altogether so plain to me as now, yet the facts were plainly indicated by Espy's storm charts.

The general views that were embodied by me in 1871, in the first edition of the little pamphlet "How to Use Weather Maps," and subsequently, in 1883, enlarged for a chapter in the proposed fourth edition have, I believe, always been nearly in accordance with the present advanced stage of thermodynamics and hydrodynamics, and are those which I have always preferred to rely upon in the work of daily weather predictions, and undoubtedly by so doing I was able to make many successful predictions (*e. g.*, the second hurricane of August, 1871) for which at that time we had little or no experience to guide us in framing empirical rules based on the study of types. These views are here now still further elaborated, and, as I hope, afford a preliminary physical basis for the philosophical prediction of storm motions, provided only the proper observations and data are at hand for any one initial weather map. Of course, however, a higher percentage of verified predictions will be attained in daily work by combining the study of types with the study of principles.

The dynamic origin of the diurnal variations of the barometer as here given is one that I have constantly urged since I first, in 1856–'60, read the famous memoir published by Ferrel in Runkle's Mathematical Monthly; but of the four sources of such variation as here explained only the second and third had at that time occurred to me, and I am still disposed to think that numerically they are the most important.

The importance of radiation as cooling the upper surface of the clouds (and of the resulting rain-fall as cooling the lower strata of air) had indeed been mentioned by earlier students but became independently and forcibly apparent very early in the course of my meteorological work and have been frequently stated by me to my colaborers. Especially was the absorption of solar heat by the upper-cloud surface and the consequent high temperature at that surface with its strong ascending currents, as described by aeronauts, persistently urged as one of the most important factors in the diurnal frequency and growth of storms; equally does the radiation from that upper-cloud surface at night cause a nocturnal maximum in cloud and in precipitation that enters as an important factor in our weather predictions. It has in fact for a long time been evident that the ascending and descending currents in the atmosphere do not constitute a strictly adiabatic process; Espy and Henry seem to have understood that the adiabatic process can form cloud only; it is the non-adiabatic processes of radiation of heat and precipitation of moisture that are needed to form and maintain storms, cold waves, etc., and when the elegant methods of Hertz and Bezold have been so developed as to enable us to solve these new complex problems elegantly and expeditiously we may hope for a careful appreciation of the disturbances thus introduced.

The great hydraulic problem of the flow of water in canals and rivers, on which so much labor has been spent during three centuries, is simulated in the horizontal movements of the atmosphere; as the mathematical problems involved have not yet been solved for a compressible gas, I have, therefore, endeavored to avail myself of the exhaustive memoir of Boussinesq in the Memoirs of the Institute of France, 1877, and of the works of Hagen and of other hydraulicians in the effort to obtain an approximate formula for the horizontal flow of a broad shallow stream of heavy air over the rough surface of the earth; this will doubtless be superseded by more rigorous general formulæ due to the pure mathematicians.

During the past year and while preparing this report I have received the elegant and important memoirs published during 1888 by Bezold, on the thermodynamics of the atmosphere, and by Oberbeck and Helmholtz, on its general motions, and have prepared for publication translations of these works. Oberbeck has, for the first time, integrated the equations of motion for fluids under conditions approximating those of the atmosphere; Helmholtz has shown the unimportance of the viscosity of the air and the importance of discontinuous motion in a train of analysis parallel to the reasoning which I had already sketched and embodied in the present paper. Bezold has given graphically the results I had attained more laboriously and numerically in 1871. While, therefore, on the one hand encouraged to think that my views can not be far wrong, I am on the other hand hopeful that these eminent mathematicians and physicists will yet give us in elaborate detail the solution of many other of the difficult problems that beset the meteorologist.

The application of numerical tables and computations to the determination of the relative forces that control the rain, the center of aspiration, the whirl, and the low pressure that characterize a storm can be advantageously replaced by graphic processes; in fact only graphic methods can completely present the resistances due to the irregularities of the terrestrial surface or the variations in buoyancy due to thickness and distribution of clouds; I have, therefore, sought for planimetric or equivalent methods of integration, but at the present moment must content myself with rapid estimates so long as direct observations of the fundamental data are so rare.

During the years 1871–'73, having in mind the train of thoughts that pervades this report, I prepared from day to day a series of measurements of the progress of storm-centers and their dimensions with the view to determining both their diurnal periodicity and the local conditions that favored their formation. The resulting data have some value in checking theoretical conclusions, and are, therefore, given at the close of this report.

In general, it is evident that the progress of the past thirty years in every branch of physical science must now soon produce a natural reaction on meteorology and enable it to take a high stand among the exact and mechanical sciences.

CLEVELAND ABBE,
Professor of Meteorology.

WASHINGTON CITY, *June* 30, 1889.

CHAPTER I.

MATTERS THAT ARE IMPORTANT VERSUS THOSE THAT ARE UNIMPORTANT.

In studying the origin and development of storms we shall need to take account of the following primary sources of energy:

1. The rotation of the earth on its axis.
2. The attraction of gravitation of the earth.
3. The heat received by radiation from the sun.

We shall also need to consider the following methods by which energy is lost, transferred, or transformed:

4. The absorption of solar radiation by the earth and air.
5. Radiation and convection of heat from the earth to the atmosphere.
6. Radiation and convection of heat from the clouds to the atmosphere.
7. Radiation of heat from the atmosphere into space.
8. The absorption of heat rendered latent in evaporation of moisture at the surface of the ground, and of the clouds, and within foggy air.
9. Resistance of the earth's surface to the motion of the air additional to those resistances that a smooth surface would offer to the steady motion of a viscous fluid.
10. The viscous resistance of the air to its own internal or differential motions.
11. The resistance, or conversion into thermal or other forms of potential energy, of the kinetic energy annulled when two currents of air interfere with each other.
12. The evolution of latent heat by the cooling and condensation of aqueous vapor.

Besides the preceding there are numerous forces and transformations that have been suggested and urged as being of possible importance; these may be considered in future more refined studies, but at the present time it is only necessary to appreciate these at their proper value in order to feel assured that we do right to dismiss them from further consideration at present. I will, therefore, make only brief mention of some of them.

(1) INFLUENCE OF LUNAR HEAT.

There is no reason to deny that the moon may have a slight influence on our atmosphere by the reflection and radiation of the heat that it receives from the sun. The amount of heat *directly reflected* at full moon is a fraction of the solar radiation less than the fraction that determines the brightness of the moon, and is scarcely the millionth part of the solar radiation. The heat *radiated* from the warmed surface of the moon must be a maximum shortly after the full moon; it is largely composed of the long waves that, according to Langley, most easily penetrate the atmosphere, therefore its co-efficient of absorption in passing through the atmosphere is on the average much less or its penetrability greater than that for the solar rays: it is, therefore, reasonable to undertake to measure the lunar radiation by apparatus at the earth's surface, such as the thermo-electric pile. But all attempts hitherto made show that the radiated lunar heat we receive is barely appreciable to the most delicate instruments, and is not the hundred-thousandth part of that received from the sun.

Since writing the preceding, Prof. C. C. Hutchins has published his determination of the heat radiated plus the heat reflected $\frac{1}{184580}$ and the co-efficient of penetrability 0.8925 while that for solar radiation is 0.75 or less.

(2) INFLUENCE OF THE HEAT OF THE STARS.

Notwithstanding the fact that the celestial hemisphere above any locality is studded with innumerable stars instead of with one moon, yet the sum total of the aggregate seems also to be inappreciable.

(3) INFLUENCE OF INTERIOR HEAT OF THE EARTH.

Observations of the temperature of the earth's surface show that heat is being conducted upwards from a warm interior, but they also show that the amount of heat thus received over the whole surface of land and water is not more than 0.0002 of all that is received from the sun directly.

(4) INFLUENCE OF THE HEAT FROM THE SHOOTING STARS.

The meteoric bodies that are continually striking upon the atmosphere penetrate it to a short distance, are heated by impact, are burned up, and disappear as shooting stars. The amount of heat evolved in such impact is easily calculated for any assumed value of the mass and velocity of the meteor, and such computations show that the total heat thus received does not exceed 0.00001 of that received from the sun directly; moreover, this heat is confined to the higher strata, whence it is finally radiated into space, so that its direct effect upon the climate at the surface of the globe is inappreciable.

(5) VARIATIONS OF SOLAR RADIATION.

There is some evidence that the amount of heat received from the sun varies irregularly owing to changes going on at the solar surface. The amount of this variation in the radiation proper has not yet been determined by direct observations, but variations of temperature have been observed on the earth's surface that may plausibly be attributed to variations in the solar radiation. Such variations in terrestrial temperatures, however, correspond to variations in the solar radiation, not greater than 0.01 at the maximum, and are ordinarily much less; they are, therefore, at present negligible.

(6) INFLUENCE OF ATMOSPHERIC ELECTRICITY.

Actual measurements of electrical potential would seem to show that two masses of air in extreme conditions may attract or repel each other electrically to an extent sufficient to produce appreciable phenomena of motion even in comparison with the far more important motions produced by solar heat and terrestrial gravity. But on the one hand such large electric disturbances as occur in thunder-storms are rapidly and immediately quieted by the lightning flash, and their energy is represented by the noise and the new nitrogen compound; on the other hand the motions produced by electrical attractions and repulsions are very local, temporary, and unimportant in the development of a storm. Even if we allow that the condensation of smaller cloud particles into large rain drops and their consequent fall to the ground depends upon the electrical discharge, yet this assumption, if adopted, will merely modify our mechanical views somewhat as follows: The latent heat evolved in condensation must be considered as not wholly consumed in directly warming the air but as partially employed in maintaining a state of electrical disturbance or tension, which latter comes to an end as soon as the flash or the silent discharge of electricity occurs; at this moment, therefore, on the one hand, larger drops are formed and fall to the ground, and, on the other hand, the energy that had been potentially present in the electric phenomena now becomes heat and warms and expands the air; thus the electric tension and its concluding flash have merely served to delay the communication to the air of the heat that was a few minutes before present in the vapor. In any event the thermal energy represented by the flash is but a small fraction of all that is present in the atmosphere as heat.

(7) THE TIDAL INFLUENCE OF THE SUN AND MOON.

The fact that the moon and sun contribute to form periodic tides in the waters of the ocean does not render it necessary to conclude that there must be appreciable tides in the atmosphere. The luni-solar tides in mid-ocean cause a rise and fall of but two or three feet at the most; the tides that are experienced in harbors are due to a combination of movements in shallow water, such as have no counterpart in the immense atmospheric ocean. The amount of probable diurnal tide in the atmosphere due to the moon was satisfactorily shown by Laplace to be not more than a tenth of a millimeter of pressure under the most favorable circumstances. On the other hand it must be confessed that his study and those of subsequent mathematicians have relation to the atmosphere considered as a quiet mass revolving with the earth: the problem of the tides in an atmosphere that is in full motion, especially the tidal actions in the northeast trades or the southwest upper currents, have not yet been mathematically analyzed; it is, however, so far as I can see, not likely that the tides as modified by such atmospheric motions can be any larger than those in a quiet air unless the atmospheric motions happen to have periodicities commensurate with those of the earth and moon, which will rarely be the case, but when it does happen the lunar tides will become temporarily as large as the configuration of the continents may allow.

Beside the possible diurnal tides which, as we have said, are inappreciable to the barometer, there is, as was pointed out by A. Poincaré, a fortnightly action of the moon that may be called a tide, which is of the following nature: Owing to the inclination of the moon's orbit to the earth's equatorial plane, which inclination varies between 18° and 29°, that satellite is during each sidereal month alternately north and south of our equator by an amount equal to the inclination of its orbit at that time. Now meteorological observations show that north and south of the equator there are belts of high pressure, and, therefore, of atmospheric accumulations; the moon's action upon these must be such as to draw the northern belt southward when the moon is south of the equator or to draw the southern belt northward, two weeks later, when the moon is north of the equator. There is, therefore, a fortnightly tidal displacement in the latitudes of the zones of calms, the amount of this displacement has not yet been determined, but is undoubtedly quite slight as compared with other sources of irregularity. Again, these belts of calms are not continuous around the earth, but are broken up by the distribution of oceans and continents; therefore the continental and oceanic areas of high barometer become as it were special objects of attraction to the moon. The locations of these belts are shown day by day in such International Charts as those published by the Signal Service for the Northern Hemisphere, and Poincaré states, a study of these shows that the breaks in the regions of calms move from west to east following the moon in its orbital motion through the month very much as the tidal protuberances in the ocean attempt to follow the moon while the earth rapidly rotates beneath. With regard to these lunar tidal inferences, it is sufficient to say that they are of an exceedingly minute order of magnitude.

Solar tidal effects must also exist comparable with those of the moon, but these are wholly obscured by the effects of solar heat which follow the same diurnal periodicity, and are, therefore, inseparable from the solar tidal effect. In fact the effects of solar heat would also wholly obscure the moon's tidal effect were it not that the period of the former is twenty-four hours or one day of mean time, while the periodic time of the lunar tidal effect is twenty-four hours and fifty minutes, so that in the course of the month the lunar effect separates out from the combined solar tidal and heat effects.

(8) MAGNETIC INFLUENCES.

As it has been claimed that there is a connection between phenomena of terrestrial magnetism and those of meteorology, and as I shall consider this connection as proper to be classed with the negligible phenomena, really unimportant in the development of storms, or general atmospheric changes, it is proper for me to briefly state here some conclusions that I have arrived at, and that I have indeed verbally communicated to several physicists. The reasons for these conclusions will be given at some future time.

According to my view the solar and lunar gravitations produce, not only the viscous tides below the crust of the earth, and the heat that warms its interior, as discussed by Darwin, Davison, and others, but also diurnal, semi-diurnal, and monthly waves of strain in the crust itself. That is to say, the actual tidal deformations of the terrestial spheroid do not wholly correspond to the possible effects of long continued steady action of the outside forces, but only partially so because of the rapid diurnal rotation of the earth with its plastic crust; the rest of their effect is a molecular straining that, like the tidal effects, goes on through positive and negative changes twice daily, having a maximum in the torrid zone, but diminishing in amplitude toward the arctic circle, and becoming positive and negative during six months alternately around the poles. The exact amount of strain and the contour of the surfaces of equal plus or minus strain at any moment within the crust proper depends upon the relative positions of the sun and moon, and the terrestrial equator, and also on the irregularities in thickness and structure of the earth crust, so that oceans, mountain chains, and plateaus will have a decided influence unless the crust is abnormally rigid; I do not say abnormally thick because the rigidity of a non-homogeneous body so large as the earth depends not upon thickness, but on the relative local rigidities of the material.

Considering the earth's crust as approximately homogeneous, and dealing only with the average strain, we must recognize the fact that at any moment the sun, for instance, tends to deform the earth's equatorial section from the circle into the ellipse as shown in Fig. 1, consequently the quadrants *A* and *B*, to which the sun is respectively zenithal and antipodal, experience a temporary strain which at the surface is one of extension, but beneath the surface is one of compression, and the quadrants *C* and *D* are in opposite conditions of strain. Each point of the equator goes through this series of alternating strains in twenty-four hours, while the polar regions *P* and *P* remain in a nearly permanent condition of strain of extension similar to *C* and *D*.

Now every solid exhibits electric phenomena when strained; some act as magnetic and others as dia-magnetic bodies; a given body when compressed, for instance, will show plus electricity at the points under compression and minus electricity at the points of its surface where extension takes place, therefore in our globe such internal electricity accumulates in the regions *A*, *B*, *C*, *D*, *P*, and *P′*, and that which is in the region *A* and *B* is either positive or negative according to the electric behavior of the constituents of our globe; quartz, and the preponderating terrestrial substances give plus electricity for compression. Hence electric currents flow from *A* and *B* toward *C*, *D*, *P* and *P′*, and as every point of the earth's surface passes around the earth's axis in its diurnal circle it comes to be alternately within a polar or equatorial current or one intermediate between them. Such electric currents affect the magnetic needle, and so far as I have examined suffice to explain qualitatively most of the peculiarities of its diurnal and annular movements.

The strained and electrified crust being a good conductor seeks by this flow of electricity to come to a condition of electric equilibrium; if the positive electricity of *A* and *B* is overbalanced by the minus electricity of *C*, *D*, and *P* and *P′*, then on the average the earth is negatively electrified; this by induction on the atmosphere gives it the positive electricity of its lower layers and the relatively negative electricity of its upper regions. By the convection of ascending and descending atmospheric currents, the positively and negatively electrified masses are intermixed in the atmosphere and lightning results in the lower layers or auroras at higher elevations.

Now the temporary electric condition communicated to the crust by this current of piezo-electricity can, through the agency of the innumerable shocks incident to earthquakes and faulting, bring the crust into a state of almost complete saturation with subpermanent magnetism. As a stroke on a soft iron bar sets its magnetism induced by the earth's magnetism, so the strokes, shocks, and vibrations incident to earthquakes, have set the magnetism induced in the rocks of the crust by our electric current. The original current or its direct magnetic effect is feeble, and, as shown by the diurnal variations of the magnetic needle, is not one thousandth part of the total magnetic intensity, but it and earthquake shocks have been coworking continuously for ages, and the result has been that the crust has attained almost the maximum amount of saturation that it is capable of. This slow secular increase of magnetism is subject to change, however, whenever new shocks occur, since they will either increase or decrease the existing magnetism, depending upon the moment when they occur; thus a region which is at one hour of the day so magnetized temporarily by the prevailing current that Airy's red magnetism is imparted to the north end of each separate mass will, a few hours after, be in such a location that Airy's blue magnetism would be imparted to the north end, so that the character of the magnetism will depend upon the time of day when the shocks occur.

The present distribution of permanent magnetism on or within the earth results from the fact that during past ages earthquake shocks and faultings have occurred by preference at certain times and positions of the earth with reference to the sun and moon, namely, at those times when the sun and moon have combined to produce maximum strains in the crust (the spring tide and neap tide of strain), and again at those times when the diurnal rotation of the earth's surface has brought the weakest parts of the earth's crust under the influence of the greatest tidal strains. For example, the Rocky Mountain range, with its continuations, the Andes to the southeast and the Asiatic range to the northwest, and the volcanic ranges of Japan and Java surround the Pacific Ocean, forming a great circle of weakness, dividing the dry land of the globe from its great oceanic hemisphere; along this great circle the crust, *i. e.*, the rigid exterior, is peculiarly liable to give way under this tidal strain, so that for ages past the mountains, continents, and plateaus of one-half the globe have been rising, the upraised strata have been plicating and crushing, the Pacific bed has been falling, and the terrestrial magnetism has by means of the attending shocks been accumulating in intensity. These changes of elevation and depression have by preference been occurring, as is shown by our records of earthquakes and eruptions, most frequently at the times when the conjunctions of the sun and moon have occurred, and especially at lunar and solar perigee and when the central Pacific Ocean has been normally under them; or in other words near midnight, Greenwich time, near the days of new and full moons during the winter months. Thus the shapes of our continents, the trends of shores and mountains, and the stratigraphic deformations depend on the positions and the actions of sun and moon upon the non-homogeneous terrestrial crust; having once, long ages ago, started an initial crack around the crust in the belt of the greatest weakness, the tidal strains have steadily increased the dislocation around this crack and have determined the times, places, and directions in which all subsequent displacements have occurred.

According to this view the location of the principal permanent magnetic influence, producing what we call the declination and inclination of the magnetic needle and the intensity of the earth's magnetism, is at a short distance below the surface of the earth, and is not distributed throughout the whole interior of the earth; it can not exist at depths where faulting does not exist, that is to say, at depths greater than 20 or 30 miles, since at these depths the great pressure welds the rocks together into a viscous mass and obliterates all faultings, and prevents the shocks needed to "set" the magnetism. At 30 miles and greater depths there may exist viscous strains and motions which will produce heat, and it is here in fact that we find the origin of the internal heat of the earth. The earth was always as plastic as now (it need not ever to have been molten in order to have become spheroidal); its internal heat is not a remnant from past cooling, but it is the conversion now daily going on of gravitation into internal molecular motions giving rise to a daily conversion in its interior of an almost inappreciable fraction of the energy of rotation into heat. This heat can not escape outward except by slow conduction, for in going downward it meets that which has come from the other side of the earth, thus maintaining the nucleus at a nearly uniform temperature, but in moving upward its progress is so slow that a considerable accumulation of temperature prevails in the viscous layer.

Now, we may assume that there is a fairly stationary condition of temperature in each layer at the present time, in other words, that the flow of heat conducted through the upper strata and lost through the atmosphere into space is just balanced by the amount produced daily in the viscous layer; but the amount thus conducted and lost is approximately known as forty-one calories per year per square centimeter, consequently we may say that the daily conversion of energy, through viscosity within the interior of the earth, is just sufficient for the melting of a layer of ice one-half millimeter thick per annum. This small quantity, an infinitesmal percentage, is annually abstracted from the energy with which the earth rotates around its axis and with which the moon revolves about the earth and the earth about the sun, and contributes slightly to what is called the secular variation of the mean motion of the moon.

These views assume that every portion of the solid crust of the earth has a certain amount of subpermanent magnetism now accumulated within it; this amount is very large; thus Gauss, on the assumption that it was equally distributed through the entire mass of the earth, determined its amount to be the equivalent of seven 1-pound steel magnets per cubic meter of earth, but these figures become an hundred times larger if the magnetism is confined to the solid crust. At once it becomes a matter of surprise that we can scarcely prove the existence of any magnetism in any of the rocks that we deal with at the surface, nor even when we go down into deep mines, as was experimentally ascertained by Carl Barus; the explanation of this, however, is simply that the magnetic condition existing in the strained rocks below the surface disappears immediately when the strain is removed, precisely as in our laboratory experiments on strained bodies, therefore the rock that we meet with at the surface and those that are freed from strain in mining operations have all lost such magnetism as they may have had before.

The above-given explanation of the nature, and estimate of the quantity, of the force implied in terrestrial magnetic phenomena shows that the meteorologist may dismiss from consideration the permanent magnetic phenomena, and the only remaining question can be as to the significance and importance of the temporary magnetic phenomena known as magnetic storms. Evidence has been adduced to show the general coincidence of such storms with the times of sun-spot phenomena. The only mechanism suggested that seems at all rational has been the suggestion that violent chemical or mechanical actions cause waves of electric influence to emanate from the sun at the times of sun-spot phenomena which, passing by the earth, cause the latter to be in its annular orbit, cutting through or intersecting a series of surfaces of equal potential thereby causing special currents on the earth; to this view there is only this plausible objection, viz, that our magnetic storms show an intensity of current vastly superior to that which is represented by the diurnal fluctuation and show, moreover, changes of intensity and character of most extravagant nature, such as we can not think likely to emanate from an immense body like the sun, whose light and heat are so eminently uniform.

Balfour Stewart has maintained the view that the magnetic fluctuations must be due to the rapid fluctuations occurring in the motions or temperature and moisture of our atmosphere as affecting the flow of electricity through it, but such storms occur in what appears to be most equable weather, and to me their explanation is more plausibly found in the frequent seismic changes going on in the earth beneath us; these latter by affecting the distribution of subpermanent magnetism must affect the closed electrical circuits or the flow of electricity through the earth's surface which must also affect the induced flow in the atmosphere above us; it is probable that all solar radiation (heat, light, electric and active) fluctuates with the solar spots and with the solar hemisphere presented to the earth so that these may also contribute slight periodical phenomena. The spots are only a visible index of what is going on at the sun's surface, but not a direct cause of terrestrial phenomena.

The magnetic phenomena in the atmosphere and the earth affect the movements of the air, if at all, then only by quantities less than the 0.00001 of those due to the thermal effects of the sun's radiation, and are, therefore, negligible at present.

CHAPTER II.

FRICTION AND VORTEX MOTIONS.

1. In studying and explaining the motions of the atmosphere it is customary to speak of the resistance offered by the ground, by mountain ranges, etc.: in order to diminish any uncertainty or indefiniteness in our minds as to what is the nature of this resistance, and, in fact, of all the resistances that affect the movement of the air, we shall consider them at some detail.

In treating of the mechanics of moving masses of solid matter the term "sliding friction" is used to indicate the resistance experienced when one smooth surface of a solid glides over another; this resistance is appreciable no matter how smooth such surfaces are; its amount depends upon the nature of the surfaces and the force with which they are pressed together. When two such surfaces roll on each other, instead of sliding, the resistance is called "rolling friction." and if we analyze this minutely it is found to be made up of two components, a normal resistance or pressure causing deformations in the yielding surfaces, and a tangential resistance or sliding friction. When an unguent is interposed between two sliding surfaces the friction is ordinarily greatly diminished, but is still often called "the sliding friction between lubricated solids," although it may easily happen that the former sliding friction is now wholly replaced by an entirely different class of resistances, namely, (1) the viscosity of the unguent or the resistance that its own molecules offers to sliding motions among themselves, and (2) the slip of the unguent, viz, the attachment of the unguent to either of the solids, and its resistance to the slipping of its own molecules along their surfaces. When the *co-efficient of slip* is zero there is no resistance to the gliding of the fluid on the solid, or there is no "external fluid friction;" when the *co-efficient of viscosity* is zero there is no resistance to the gliding of the molecules of a fluid on themselves, or there is no "internal fluid friction;" when the *co-efficient of sliding friction* is zero there is no resistance to the sliding of one surface of a solid on another.

2. As the word friction has been used in a generalized sense in mechanics to express indifferently any form of resistance tangential or normal, massive or molecular, so an indefinite use of the word has sprung up in meteorology to indicate any sort of resistance to atmospheric motions. The movement of the atmosphere near the earth's surface is commonly said to be greatly affected by friction, and this is said to be

illustrated by the fact that often calms or light winds prevail below while clouds, balloons, and smoke show a strong wind above the calms. Such illustrations are confusing. When a horizontal current of wind occurs at some distance above the earth its action upon the atmosphere below is determined by these considerations, viz:

(*a*) If the upper current is neither lighter nor denser than consistent with the lower air, that is to say, is in hydrostatic equilibrium, it will remain at its own elevation, moving horizontally, gliding over the lower air, and acting upon it only by virtue of the sliding friction of air upon air, the proper name of which action is "internal friction of gases" or "viscosity of fluids"; but in the case of air this viscosity is so slight a resistance that layers of air a few feet apart or even a few inches apart and even when in rapid motion may be considered as moving independently of each other; let there be a deep stratum of air of ordinary density and at a temperature of 0° centigrade, whose upper surface has a given horizontal movement suddenly imparted to it, then it will require twenty-six years to transmit to a layer 100 meters below this surface a velocity one half of that which prevails above it if the viscosity of the air be the only means of effecting this transmission.

(*b*) When the rough surfaces of two solids slide upon each other it is conceivable that their slight elevations and depressions interfere with each other, and that the shocks thus produced consume a portion of the energy of the moving masses, thus apparently increasing the so-called sliding friction; applying this idea to the atmosphere many have written as though the co-existence of calms below with winds above is in some way due to the irregularities on the earth's surface by which, for instance, the air in a valley is entirely hindered from moving while the wind blows freely above the mountain tops, but this is a very unsatisfactory explanation of the phenomena as will be evident on considering that the air over prairies, plateaus, and even the ocean, where no mountains exist, is frequently calm during the night time while strong winds are prevailing over head; and on the other hand the air in the valley moves during the day time even though it be calm during the night; the impact of horizontal currents against the irregular surface of the earth, or the so-called resistance of the earth's surface, although it must retard the motions yet can not be the sole cause of the general phenomena of calms and light winds that often prevail during the night time and during cloudy or foggy weather while the wind is blowing overhead.

(*c*) The important fact that is always to be kept in mind is that the moving upper air will not descend to the earth and set the lower air in motion unless the upper layers are abnormally dense or the lower layers abnormally light so as by their relative densities to cause a vertical interchange of their masses. Such conditions as to density do not generally exist at night time and exist in only a moderate degree during cloudy and foggy weather, but they are common enough during clear weather in the day time; therefore instead of saying that nocturnal calms at the earth's surface are due to the great friction between the earth and the air, or between the upper air and the lower air, one should say that they are due to the absence of any cause by reason of which the upper air should descend and communicate its horizontal motion to the air below; similar remarks would apply to calms on the leeward side of a mountain or building; for this reason also the contrast between the night wind and day wind is greatest over the hot continents, and least over the cool, moist ocean.

(*d*) Therefore "friction" or resistances due to slipping and impact and viscosity do not entirely cause or explain the co-existence of upper currents and lower calms; it is the ascending and descending currents that exist in the day time and the impact of air against the irregularities of the earth's surface that give rise to the dissipation or transformation of energy summarized in the word "resistance," and these may be included under the general term friction (and have in fact been so included by others) provided that specific names are given in order that our expressions may have the necessary clearness and definiteness; this subject will now be further considered.

3. When by its impact on any obstacle the horizontal movement of the air is converted into the curvilinear or vorticose motions, and the energy of its direct motion thus diverted in other directions, the resistances thus introduced by objects external to the moving air are said to be due to external fluid friction as distinct from the internal friction or viscosity. The motions thus introduced into a fluid mass consist of sudden changes of direction of an irregular kind by which smaller masses of fluid are made to impinge on and push against each other, or move away from each other as the case may be; there are also more regular curvilinear motions such as whirlpools, or other vortices; even vacuous spaces may be formed in the water within a whirl or on the lee side of obstacles, while in the case of air regions of small density approaching a vacuum may be formed. Thus the motion of the whole fluid instead of being in parallel lines and at uniform velocities, as in permanent steady motions, is broken up into an assemblage of masses within each of which complicated movements are taking place; the movements within each such smaller mass may be quite independent of the motions in the neighboring masses. Between two such contiguous smaller masses or between a whirl and the outside irrotational fluid a neutral surface or layer of fluid exists which is a boundary surface between regions of what was first by Helmholtz called discontinuous motion; if on one side of such a surface of discontinuity there be a vacuous space whose vacuity is due to the motions within the surrounding fluid such region is called a discontinuous space. The word discontinuous in this connection is used to indicate the fact that if there were no viscosity then there would be no necessary connection between the motions in two contiguous fluid masses separated by such a boundary surface. The ideal perfect fluid is very nearly realized in the atmos-

phere so that in it a large class of rapid motions take place in close proximity to each other without any material connection.

In the accompanying Fig. 2, *a*, *b*, *c*, *d*, is a boundary surface between the whirl within and the motions outside, such that if the liquid fluid be non-viscous or only slightly viscous the motions within and without the surface are entirely independent of each other, and depend only upon independent initial conditions. Therefore as one passes through the mass of fluid he passes through a region where one law of motion prevails up to a boundary surface and then suddenly into a region where entirely different motions prevail. The word discontinuous, therefore, refers not so much to a discontinuity of the motion, or the matter, as to the discontinuity of the law of motion. It is not plain how such vortex motions could originate in an ideal fluid, but the treatment by Helmholtz of the equations of motions for such a fluid shows that when once in existence such vortices must continue in motion without change and without destruction, although the movement may be transformed in various ways. Some of Helmholtz's conclusions are thus summarized:

(*a*) In a perfect fluid a vortex will always contain the same elementary fluid atom that was in it at the beginning of the motion, no matter how the vortex may move or change its shape;

(*b*) The cross section of the vortex ring multiplied by the angular velocity of that section around its central core is a constant quantity and is called the strength of the vortex;

(*c*) A vortex ring, whether circular or not, forming a closed curve of any shape whatever must always remain closed if the surrounding fluid is of indefinite extent;

(*d*) Two vortices at a distance from each other may affect each other through the influence of the intermediate fluid;

(*e*) An isolated circular vortex moves with a constant velocity along an axis perpendicular to the plane of its circular core and in the direction of the motion of the fluid on the inner surface of the ring.

Circular and especiallly cylindrical and other vortices play an important part in atmospheric movements and will be referred to further on, since a slight consideration of the motion of a mass of gas or liquid surrounded by similar fluids shows the prevalence everywhere of vortex motions.

4. A careful consideration of a few simple experiments will give precision to our ideas as to the movements that take place in our atmosphere.

(*a*) Let a tube be so arranged at the bottom of a vessel of water that a drop of oil is detached as at Fig. 3 (*a*); it assumes a flattened form (*b*) and rises to the surface by reason of its own buoyancy. If the drop be not too minute we see that as it rapidly ascends its upper surface is decidedly flatter than the lower or rear surface as at (*c*). The drop has undoubtedly assumed a shape such that the curva-

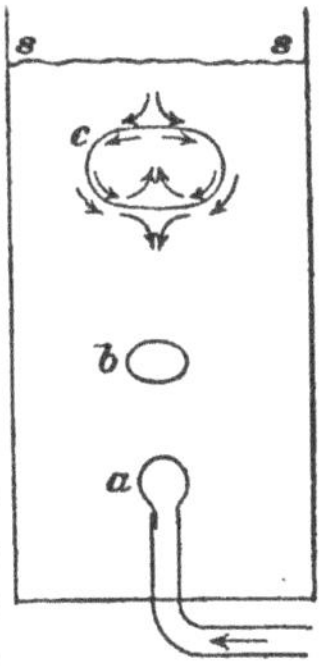

FIG. 3.

ture at any point depends not only on the surface tension for oil and water and the static pressure of the water but also on the inertia of its own motions and the viscous friction due to its motion through the water; it is, in fact, a body of least resistance for the given velocity, viscosity, and internal motions. If we examine the drop carefully we shall see that each point of its surface has a motion as shown by the arrows in Fig. 3 (c) while the surrounding water has the feebler motions also indicated by the arrow.

The rapidity of ascent of the drop increases for a short time and then becomes uniform. With drops of various sizes the rapidity of ascent increases with the size of the drop; as the buoyancy of a unit mass is the same for all sizes this shows that the upward velocity depends principally on the relation between the buoyancy and the viscosity of the two liquids; the ascent is, therefore, more rapid in warm than in cold water, other things being equal, because of the diminished viscosity of the warm water.

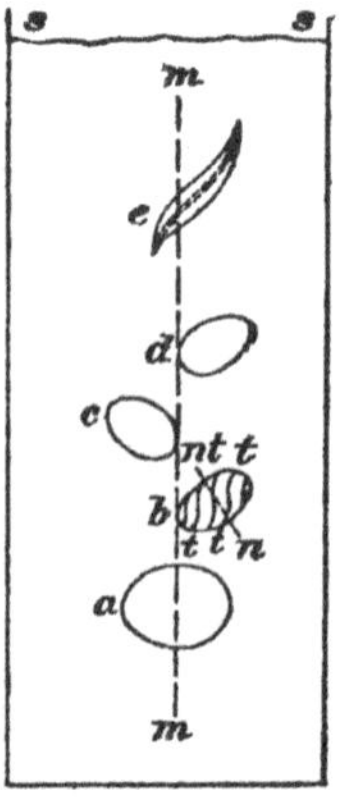

FIG. 4.

(b) If the drop of oil be still larger it seems to a single observer to rise as in Fig. 4 (a b c d); apparently the drop swings to the right and left as it rises but this is only an optical deception; the true general motion is that of a spiral ascent, the drop revolves about a vertical axis passing through it; the upper surface that was before nearly horizontal now becomes inclined; the time of rotation about the vertical axis is the same as the time of ascent between the position b and d; the motions of the surface particles are now more complicated than in Fig. 3 (c); they describe curves inclined both to the fixed vertical axis *mm* and to the normal *nn* being curves of double curvature such as *tt* in Fig. 4 (b).

(c) A further increase of the size of the ascending masses is accomplished by liberating larger masses of oil and thus the mass, Fig. 4, (b) becomes a lengthened mass (e) whose rapid ascent and internal rotation about its inclined curved dotted axis, which latter varies but little from a vertical axis, gives it a form resembling somewhat the ascending lambent flames that attend an open fire of wood or coal.

(d) If finally a uniform steady flow of oil be maintained it streams upward as a cylinder, as at 5a and 5b, with a steady rotation about the vertical axis; if the distance up to the surface of the water be sufficient this column of oil will break up before reaching the surface into separate masses and drops as at 5c; the place at which this break occurs is determined by the fact that on the one hand the rate of rotation and consequently the internal tendency to disruption by reason of centrifugal force increases steadily from 5a to a maximum at some point such as 5d, while on the other hand the flexible surface separating the oil from the water allows any accidental disturbance to be propagated as a surface wave from $5a$ to $5b$; this disturbance, compounding with the rotation, forms a fluted cylindrical vortex or a vibrating columnar vortex (see Thomson, Hicks, Basset, etc.), and accelerates the disruption.

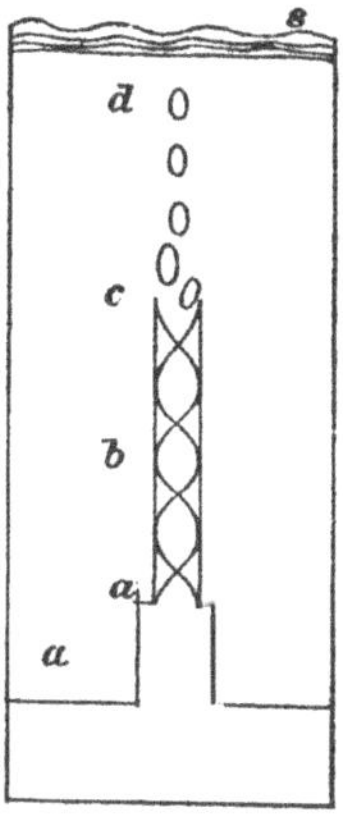

FIG. 5.

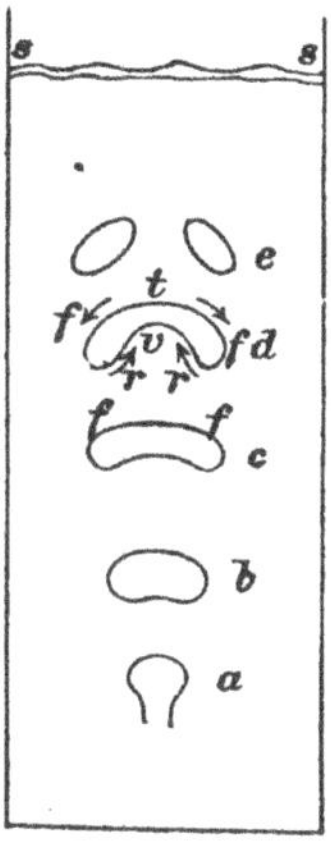

FIG. 6.

(*e*) If now we pass to the more rapid ascent of bubbles of air in water, as at *a* in Fig. 6, we note that very small bubbles ascend as flattened spheres, but large ones assume the umbelliform as at *c* in Fig. 6. The large surface tension of water and air has had an effect on these small bubbles similar to that of the feebler tension of oil and water in the preceding experiment, but the great rapidity of ascent due to the greater buoyancy of the air has produced a greater curvature of the front surface *ff* in Fig. 6 (*d*), along whose surface the particles pass from *ff* around to *rr*.

The resistance experienced by the drop as it rises is at first equal only to the force required to do the work of overcoming viscosity in the surrounding fluid since the internal viscosity of the air is so much smaller than that of water. The thickness from *a* to *b* diminishes as the bubble increases in velocity of ascent and therefore increases the rate of motion from *f* around to *r*; for a moment there is formed a complete vortex ring of air, but instantly this ring is broken into parts that break up into smaller bubbles of air, some of which may go through the same process, until finally all the air is divided into bubbles small enough to ascend steadily as did those of oil in Fig. 5.

(*f*) We now pass to the case of warm colored water rising in cooler water. Here the buoyancy is relatively smaller, being only that due to the differential expansion by temperature, and the viscosity and surface tension are also smaller, being those due to warm and cold water, and a possible influence of our coloring matter in altering the viscosity. If a slight progressive motion be given to the ejected fluid it rises by virtue of this and its own buoyancy, describes a short path as a vortex ring and comes to rest at *d*, Fig. 7; the time of stopping is determined by

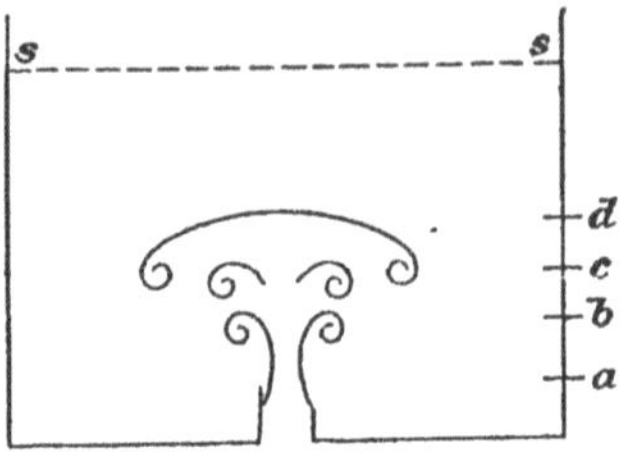

FIG. 7.

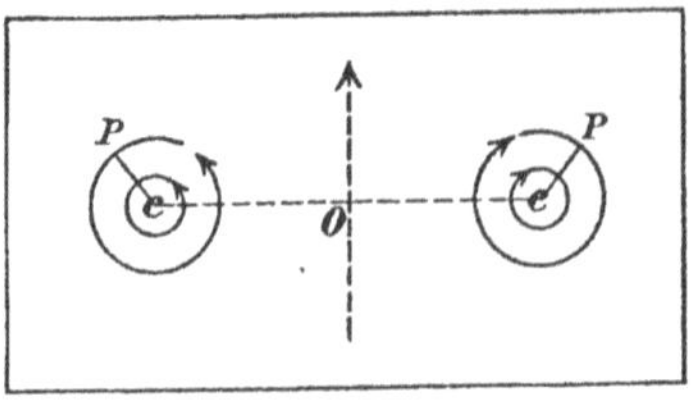

FIG 8.

the time required for work done against viscosity to consume the initial momentum of the mass and the further time required by the mass to lose by conduction of heat the buoyancy due to its temperature. Assuming the motion to be slow and that no appreciable mixture of cool and warm water takes place it is still evident that the internal motions of the vortex mass are continually bringing new layers of particles to its surface and thus a large layer of particles of the colder external water is by surface contact and conduction causing the initial heat of the mass to be rapidly abstracted; but mixture is always taking place, to a slight extent in liquids and rapidly in gases; the original mass may break up into smaller ones if its initial motion is more rapid or its water very hot, and the ascent be stopped proportionately soon by this process of differentiation.

If the initial mass has the same temperature as the surrounding fluid and has only an initial motion of translation without destructive perturbations the vortex ring moves intact until viscosity brings it to rest.

5. We now pass to air moving in free homogeneous air. We have here only the slightest effects of viscosity and no surface tension but large disturbing effects due to difference of temperatures, cooling by conduction, and dissipation of energy by mixture. If in the original mass the temperature be the same as in the surrounding air the destruction of its distinctive motion depends on the viscosity, and of course on any outside disturbing influences. In experiments with smoke rings (see Figs. 7 and 8) in air of perfectly uniform temperature, in sheltered rooms, such vortices continue to retain shape and motion until viscosity consumes their momentum, but in the open air the irregular currents and densities soon break up such vortices into smaller ones of distorted and multiform shapes until viscosity has consumed their momentum and their higher temperature if any has been communicated by conduction to the air with which they are mixed.

The velocity with which a circular vortex moves through quiet air in a direction perpendicular to the plane of the vortex ring was determined first by Sir William Thomson in 1860, and subsequently by J. J. Thomson, and others. If the air be perfectly free from viscosity then the velocity of progress of the ring is uniform and stable and is given by the formula (see J. J. Thomson, Motion of Vortex Rings, p. 33)

$$V=\frac{m}{2\pi a}\left(\log\frac{8a}{e}-1\right)$$

where a is the radius Oc of the ring in Fig. 8, or the radius of the axis of the core.

e is the radius cP of the circular section of the ring, which is supposed to be of uniform section.

ρ is the density of the gas, supposed to be the same within and outside of the ring.

α is the uniform angular velocity with which the radius cP and every particle lying in any section of the ring revolves about the centre c.

$m=\pi e^2\alpha=$ the "strength of the vortex," or the product of the angular velocity by the area of the section of the circular ring.

In the case of viscous fluids the movement of the vortex ring is not stable but is continually being destroyed by the viscosity which causes portions of the external quiet fluid to be brought into the vertical motion, and portions of the vortex to be left behind in the quiet fluid, the result of which is that the ring is continually enlarging its diameter Oc, and also its radius cP, and is, therefore, moving more and more slowly until finally brought to rest. The law according to which this viscous effect is developed has not been given deductively but is approximately known from experiments; thus R. S. Ball, in 1871, found for rings of 9 inches initial diameter and for initial velocities up to 10 feet per second, that they describe a length of path before being brought to rest of from 2 to 20 feet. The rate of diminution of the velocity of the ring showed that its surface was experiencing a resistance that varied as the velocity, not as the square of the velocity, and, therefore, a resistance that exactly comported with the laws of viscosity. The exact rate of retardation per second at any moment for these rings was 0.37 of the momentary velocity expressed in feet per second.

6. When, by the ascent of small cylindrical or circular vortices of warm air from the surface of the ground to a height of several hundred feet, the lower atmosphere has become thoroughly mixed and very slightly warmer than is consistent with stable equilibrium, there then begins the formation of ascending currents on a still larger scale; these represent the warm air from regions on the earth's surface that are specially liable to become overheated and there begins a slow steady ascent which is most likely to be accompanied by a slight rotary motion, so that quite large cylindrical vortices are formed, invisible to the eye and of a temperature only very slightly greater than of the surrounding air. The rate of ascent of a particle near the surface of such a cylindrical vortex for perfect fluids is known by the studies of Rayleigh on the theory of the so-called liquid jets, namely, those without viscosity (see Bassett, Hydrodynamics, Vol. II, pp. 191 and 315), that applies to the lower part of the column shown in Fig. 7, namely, from a to b, above which the cylinder breaks up into separate circular vortex

rings. Such breakage is antagonized and delayed by the actions of viscosity. In our own atmosphere, and for the columns that feed the clouds, such breakage will on a very dry day occur at heights of from 1 to 3,000 feet, and the separate masses above *b* slowly rising and floating with the horizontal wind continue invisible until, as in Fig. 31, page 72, they rise to the level of cloud formation and begin to form separate small cumulus clouds at *K* after having cooled by expansion to the dew-point.

7. Owing to the inertia developed by the centrifugal force attending the rotations within circular and cylindrical vortices, and no matter whether the axes are vertical or horizontal, or in any other direction, the elastic or static pressure within such vortex is less than in the quiet fluid outside. The pressure of the latter against the boundary surface of the vortex is balanced by the pressure within the vortex, due to its static pressure plus that due to its centrifugal force of rotation; a small portion of the energy within the vortex is through viscosity converted into the elastic static pressure due to heat in the contained fluid, so that the pressure within that fluid is slightly increased in proportion to the strength of the vortex. In viscous fluids all the energy of the vortex motion, which is simply the energy of the buoyant ascent, is eventually converted into heat and disseminated through the resulting quiet fluid. This is, however, but a small effect. On the other hand, if the velocity of rotation exceeds a certain limit the inward pressure of the outside fluid is not sufficient to counterbalance the outward pressure due to centrifugal force and unless, as in liquids, the surface tension assists in restraining the expansion of the vortex it will follow that, as in the case of gases, the vortex at once enlarges its radius and diminishes its centrifugal force so that it is able to balance the free static pressure on the outside, but on the inside it maintains a hollow core that is vacuous if we are dealing with viscous fluids—liquids that have the property of surface tension, or partially vacuous, if we are dealing with gases; such cores of rarified air are frequently seen in small whirlwinds and in larger tornadoes.

The conditions under which such hollow cores are prevented in circular vortices is expressed by saying that the elastic pressure Π in the fluid surrounding the vortex must be greater than (see Bassett, Hydrodynamics, Vol. II, page 87)

$$\frac{\mu^2\rho+\mu'^2\sigma}{32a^2b^2}$$

Where

$\mu=\frac{\pi}{a}$=circulation of the outside fluid

ρ=density of the outside fluid

$\mu'=4m\pi a^3b^2$=circulation within the circular ring vortex

σ=density within the circular ring vortex

$a=Oc$ in Fig. 8

$b=cP$ in Fig. 8.

In general the pressure within the vortex will be equal to the pressure outside the vortex, diminished by that due to centrifugal force, or (Bassett, Vol. II, p. 86, eq. 83)

$$P=\Pi-\frac{\mu^2\rho+\mu'^2\sigma}{32a^2b^2}$$

8. It is believed that the processes taking place in the atmosphere are not very dissimilar to those that can be reproduced in air or even in water in the laboratory. Thus, let Fig. 9 represent a jar of warm water with which I had occasion to make a few observations with a view to elucidate this subject, the surface of the water being at a temperature of 90° Fah., the air temperature 70°, the dew-point of the room about 60°; the mass of vapor-laden air streaming over and up from the water assumes every variety of motion, rectilinear steady streams, rectilinear vortices, cylindrical vortices, vortex sheets, circular vortices, fluted vortices both circular and cylindrical, even interlocking circular vortices. These all disappear to the vision within a few inches of the surface because the vapor that made them visible is rapidly mixed with the drier surrounding air that is moving past with the currents of the room and descending through the warmer and lighter steam. This mixing is a process of diffusion and of convection, the latter being seen in the process of rolling and unrolling that is rapidly going on in the vortices.

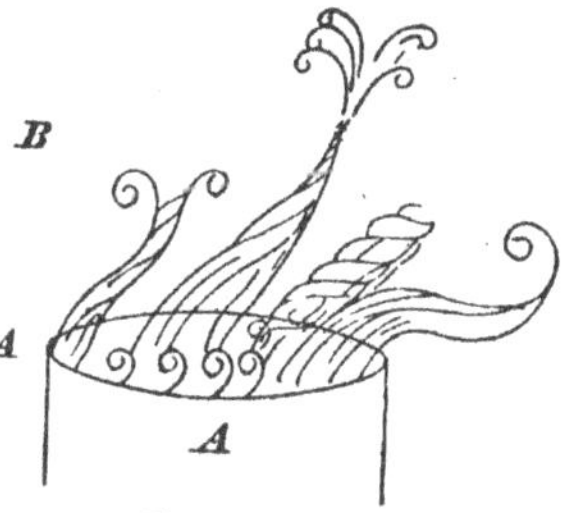

FIG. 9.

Thus the air just above this layer of visible vapor as at *B*, Fig. 10, comes to assume a temperature higher than that of the air in its immediate neighborhood, and higher than it had previously possessed; it also has a dew-point higher than before. The excess of temperature and dew-point of the moist air within over that outside the column *B*, gives it an excess of buoyancy ($\rho-\rho'$) nearly proportional to the weight of warm vapor that is mixed with the previously drier and cooler air. Thus it happens that above the surface of the warm water there is a large mass of lighter air, the appearance being somewhat as in Fig. 10, where we see the visible vapor at the bottom of a fluted column in which it rises and becomes invisible, but to which it contributes the buoyancy of its own heat and lightness. This ascending column inevit-

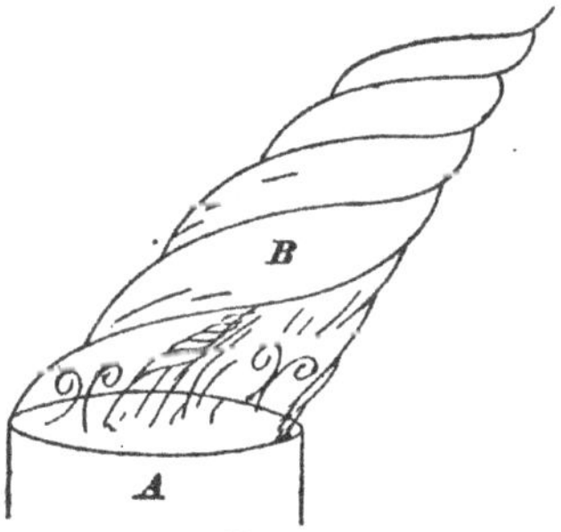

FIG. 10.

ably acquires a rotary motion which is, however, slower than that of the original vapor streams in Fig. 9; its rate of ascent is also slower. For small columns, the rates of ascent and of rotation are principally controlled by the average buoyancy of the air comprising the column. As shown in this diagram the fluting that marks the column indicates the beginning of internal and differential motions that will eventually disrupt the ascending mass, no matter how carefully we attempt to shield it from outside disturbances. The height to which a fluted column will ascend before disruption depends upon the horizontal diameter, the buoyancy or velocity of ascent, and the initial disturbing action.

9. In the small cylindrical and conical vortices of Fig. 9 the height was from three to five times the diameter of the base. In other vortices, such as that of Fig. 10, where the buoyancy is less and the diameter greater, heights of five, ten, or fifteen times the diameter of the base are attainable. No general rule can be adopted for the free atmosphere, but a height of ten times the diameter of the base is frequently obtained, perhaps more frequently than larger or smaller ratios.

No sooner has the rising column become divided into separate masses than these each continue their ascent as individual, separate, circular vortices, as in diagrams Nos. 12 and 30, which generally rapidly lose their vortex motion by reason of atmospheric viscosity and the cooling due to conduction and radiation consequent on mixture, the result of which mixture is the formation, at an elevation of a few feet, of an amorphous mass of air possessing a slight buoyancy; the latter in its turn may begin a still slower ascending motion, finally ending in the formation of a higher amorphous layer by the mixture of it with similar large columns, and thus heat and vapor slowly ascend toward the level where cloud formation begins.

The formation of clouds, both of smoke and of vapor, offers us daily examples of the vortex motions, the mixtures and the successive layers that are present in the atmosphere; thus, in Fig. 11, we see at *A* a rapidly ascending column of great buoyancy and moisture, by the rapidity of its ascent and rotation it is the more quickly broken into elementary vortex rings as at *B*, these rise and spread as at *cc*, while the central vortex column supplies a steady influx of lighter air, so that from *B* to *D* we have a series of rapidly ascending, expanding vortex rings, each of which may circulate slowly as a whole in a nearly horizontal plane around the general axis *AB*. As the expanding rings cool, both by expansion within and by mixture with the cool air on the outside surface of the cloud, as well as by radiation, they eventually become visible as clouds of vapor. As the sun shines upon one side. *S*, of the cloud, leaving the other side, *X*, in darkness, the vapor on the side *S* becomes warmer and lighter, and therefore rises higher than that on the side *X*. Some, especially the outer rings, breaking up into smaller divisions, perhaps even losing their buoyancy, descend, after having risen by their inertia higher than they were able to stay, so that the general growth of the whole, considered as a mass of vortices, may be described as a lower jet, *AB*, breaking up into a mass of rising vortices in the central portions of the cloud and of descending vortices in the lower outer portions such as *Y* and *Y*; when these latter reach a plane at which the pressure is nearly the same as that at *B* the vapor is redissolved and they disappear from sight so that the cloud has a nearly horizontal limit at its lower surface, no matter whether that be defined by rising condensing vapor as at *B*, or by descending evaporating vapor as at *Z*.

FIG. 11.

The precipitation of the vapor into visible cloud and the evolution of latent heat is a reversible process, since the evaporation of that same cloud would again consume that same amount of heat, but, if any of this heat is lost from the cloud by this radiation, or is used in warming up the intermixed cool air, and a corresponding portion of the vapor is eventually precipitated to the earth as rain, snow, or hail, the process is to that extent irreversible. This precipitation will take place sooner and more easily on the shaded side of a cloud, *BXZ*, than on the sunny side, *BSD*, or rather it will take place from the shaded surface, *XZ*, rather than from *DS*, since the thickness of a few yards of cloud is sufficient to cut off most of the solar rays and of the radiation into space.

10. The heat consumed in evaporation and in heating the air at the earth's surface is thus eventually conveyed to the upper limit of the clouds, and we know that it must all be given up by radiation to the outer space, otherwise there would be a secular accumulation of heat and moisture in the atmosphere which the permanency of the climate shows

is not the case. But this convection of heat requires *time*; when it takes place more slowly than the supply is being received from the sun day by day, the lower air accumulates heat and the temperatures rise; the layer of warm air grows thicker and the lower convection currents rise more slowly (because their buoyancy is diminished, being proportional to the difference between their own temperatures and that of the surrounding air), but they rise higher because they retain their buoyancy longer; a day comes when they rise high enough to form small cumuli during midday, then as days go on they form large cumuli, then thunder-storms, water-spouts, and tornadoes.

The large masses here considered are columns of air covering a considerable region of the earth's surface, having a slow ascent, and a slow rotation, the latter being due in part to local outside disturbances, but also in part to the influence of the rotation of the earth; when the mass of air attains any considerable size, such as one mile in diameter, the earth's rotation becomes a controlling force, wherefore such columns in the northern hemisphere almost invariably rotate contrary to the hands of a watch. The mathematical and mechanical treatment of the laws of motion of such cylindrical gaseous vortices or jets have been accomplished for the case of compressible and viscous fluids like air or water, but they have only a distant similarity to the laws of motion of incompressible perfect fluids with appreciable surface tensions. The instability of motion of a cylindrical jet of water has been beautifully studied experimentally and theoretically by Bidone, Savart and Rayleigh (see Proc. Roy. Soc., London, 1879, XXIX); the conditions under which the jet breaks up into drops are thus partly known; but in a rising jet or column of air there is no surface tension to hold it together, and the motion is broken up by its resolving itself into a spreading umbelliform figure as at *d*, Fig. 7, which shape is assumed when the viscous resistance and the loss of buoyancy due to cooling have finally consumed the energy of rotation or the strength of the vortex. On this account the top of the cumulus cloud finally spreads out horizontally as the invisible vortices of the clear days preceding the cumulus days had done. Therefore the height to which the upper surface of the warming stratum in its steady daily growth will have attained at the end of any given day is determined by the height necessary for the rising jets to attain before viscosity and cooling rob them of their buoyancy.

1. In the preceding we have considered how the intrinsic buoyancy of special masses of air causes local ascending currents within a larger mass of atmosphere, which may itself be either stationary or advancing over the surface of the earth; but for every ascending mass there must be a descending mass, and on the average these masses and volumes are equal, although their densities are unequal. The laws of thermodynamics show that within such descending masses the temperature must be steadily increasing, at a rate which, for the lower portion of the atmosphere, is very nearly uniform over a large portion of the earth's surface at any one time, and varies very little from 9.5° C. for a descent of 1,000 meters, or 5° Fah. for each 1,000 feet. Notwithstanding this rapid increase in temperature during descent the air will continue to descend so long as it retains any deficiency of buoyancy, and will finally rest only in regions of the same density as itself, or, if need be, at the surface of the earth. Cool air bringing down fog or cloud will warm up more slowly, and, therefore, will descend lower than air of same initial temperature without fog. It is important to acquire some definite conception of the limiting altitudes between which this vertical interchange takes place.

2. We have seen that the maximum altitudes attained by the lower ascending currents correspond to the tops of the cumulus clouds, and for daily storm studies it will be a sufficiently close approximation to assume that the descending currents start from points not much higher than the cirrus cloud, or 15 miles at the maximum, but for seasonal climatic changes and periods we must consider even higher layers. As the summit of a cumulus cloud usually has a small horizontal section compared with that of its base so, *vice versa*, the lowest point of the descending current has, while rapidly descending, a small section compared with that at the higher elevation at which it began; we may, therefore, in general, consider that the spaces, *SSS* (Fig. 12), of blue sky between small clouds represent regions within which the air is, on the average, descending very much as shown in the figure; that the apices of the invisible descending masses as shown at *AAA* correspond inversely to the apices of the cumuli *CCC*, and of the ascending streams *RRR*. The principal difference between the apices of the ascending and de-

scending currents is: that within the clouds the ascent due to buoyancy is maintained by the latent heat of condensation of vapor, and when this supply is diminished the rate of ascent diminishes, so that the tops of the cumuli may ascend with comparatively great slowness, and may even be stationary or falling. On the other hand, the rate of descent within the invisible descending masses of denser air depends upon a deficiency of buoyancy that is unchanged as the mass descends (except by the effect of mixture, and the more important effect produced by the discontinuous motions that perpetually break up the larger masses of air into smaller ones and mix them with the surrounding air). The result is that a considerable portion of the descending or ascending mass becomes so thoroughly mixed with the mass through which it passes, that it loses its individuality and contributes a portion of its mass to form the general stratum *BB*, within which (1) the other masses, *DDD*, that have not lost their individuality continue to descend, some of which are only stopped on reaching the ground, and (2) other masses that have not lost their individuality, *i. e.*, the *AAA*, ascend to form clouds.

That such small masses of denser air are continually settling to the ground during the hotter portions of the day in clear weather will be evident to any one who watches the phenomena taking place over a dusty road or other regions where very light objects are available to show the direction of motion of the air under such circumstances. When everything has been quiet for a few minutes it is very common to observe the dust suddenly begin to fly away from a central spot on the ground with considerable violence and without apparent cause, until we see it has been blown from that central point outwards in all directions, and it is evident that a supply of air has poured steadily down upon that central region, and, spreading out on all sides, has finally flattened out over the earth's surface, replacing the very hot air that had been there a moment before by some that is cooler, drier, and denser.

The diagram, Fig. 13, shows at *a* the approximate form of such a descending mass of air, which, if undisturbed in its descent, would retain the general form of a descending vortex ring as at *aa*, until when very near the ground the front portions *bb*, become less prominent, and the central descending portion *c* strikes the ground first. Fig. 14 presents a sketch of a section in perspective of the air currents as shown by the dust raised from the ground and driven from *c* outward in all directions. Fig. 15 shows the final result, namely, a thin layer of cool air whose density is slightly greater than that of the surrounding air covering the region upon which the dust has again settled, and which will quickly warm up and be ready in its time to be displaced by a mass of denser air. It should be carefully borne in mind that the superior density of these descending currents is due not merely to lower temperature, but equally to lower dew-points or dryness.

3. The general expression for density (not specific weight but specific

mass) of air at a pressure b, temperature t centigrade and containing vapor at a tension of e is

$$\delta=0.001293052 \frac{1}{1+0.003670\,t} \frac{b-\frac{3}{8}e}{760}$$

and the values of these factors are tabulated in the physical-chemical tables of Landolt and Börnstein (Berlin, 1883) whence we see that a rise or fall of 1 degree centigrade in temperature diminishes or increases the density in the same proportion $\mp$ 0.00367 as would be produced by a rise or fall of 0.01 in the ratio $\frac{e}{b}$ or a fall or rise of 7.6^{mm} or 0.3 inch in the actual vapor tension e which corresponds to a change of 7° Fah. in the dew-point when the temperature is 85° and to larger values at low temperatures, as is easily found from any table of vapor tensions. The change of $0.00367 \times b$ in the ratio $\frac{b}{760}$, or a change of 2.79^{mm}=0.11 inch in the barometric pressure (b) produces the same proportional change of 0.00367 in the density.

If air by rising a little (Δh) decreases its pressure (Δb) its temperature (Δt) and vapor tension (Δe) simultaneously at the uniform rate of Δb, Δt, Δe, per hundered meters (Δh) then the resulting change in density ($\Delta \rho$) is found by adding the individual small effects or

$$\Delta\rho = \rho^{1}_{0}\,[+\;0.001316\Delta b + 0.00367\Delta t + 0.00050\Delta e] \times h.$$

4. It is common to say that the upper air descends to the earth because it is colder, but when the dynamic heating of compressed air is considered it is, at first thought, difficult to realize why descending air, warming up at the rate of 1° C. per 100 meters of descent, should ever be cold enough to descend to the earth's surface at all. Now, as the general layer of atmosphere *BBB*, Fig. 12, is itself only approximately in convective equilibrium a mass that is slightly denser than its surroundings, will on descending still generally find itself somewhat denser than its surroundings, and since it retains its low dew-point which is not changed by its dynamic warming or cooling, and only slightly changed by the mixing of moist air, therefore, by the time it reaches the earth's surface its deficit of buoyancy may be owing to its very much lower dew-point quite as much as to its slightly lower temperature. But this lower dew-point in air that contained much moisture when it was at the earth's surface in some earlier stage of its history, can only have been brought about by the loss of moisture due to precipitation when the air was near the summit of its vertical circulation, after which loss of moisture it then by radiation lost its heat, and only then did it become denser than its surrounding air and begin to descend. Thus all convective processes in our atmosphere resolve themselves ultimately into the descent of air that is denser because it is drier, and it is after such descent has begun that the cooling by radiation goes so far as to make the air continue denser because it then becomes colder than the surroundings.

The phenomena here noted I have often observed and utilized in connection with my predictions of thunder-storms, and have also discovered that they were known to Espy. (See Journal of Franklin Institute, Vol. VII, 1831, pp. 224 and 225.)

5. The ultimate origin of or cause for the existence of regions of abnormally dense air at the level of the upper portions of the stratum *BBBB* is found in the following three considerations:

(*a*) During the night time, in clear weather, such density is due to coldness produced either by special irregularities of radiation from dry masses of air, or in mountainous countries by the horizontal forcible transfer by the wind of the cold air from the mountain tops and plateaus; but the latter air owes its coolness to radiation from the high lands.

(*b*) During the day time any special density in localities in the upper portion of *BBB* is due to the dryness and coldness of air that has lost its moisture as rain and its heat by radiation, after ascending with the temperature and moisture of the overheated air at the earth's surface.

(*c*) The air ascending over hot, dry regions of the earth arrives at a certain high level denser than the air that has ascended from neighboring moist regions; *e. g.*, air from continents becomes cooler than air from oceans.

As density or buoyancy is by far the most important cause of uprising currents there results a marked diurnal periodicity in the formation of cumulus clouds, and with it a corresponding periodicity in the formation of the turbulent motions and whirls between the earth's surface and the tops of the cumuli. The condition of such a mixture of vortices approximates to the state discussed by Maxwell in the Philosophical Magazine, 1861, on the theory of Molecular Vortices, and differs from ordinary quiescent motion of air by the dissemination throughout the mass of a pressure depending on the centrifugal forces of the vortices. In this way we may perceive that the pressure within the mass of the lower atmosphere will increase as soon as vortex motions are formed in the otherwise steady irrotational motion of the ascending and descending currents. The regions where the vortices are of smaller dimensions and more numerous are those in which they revolve most rapidly, and consequently may even be those of greater rise in pressure. But vortex motion is destroyed by viscosity and the energy of the moving mass is converted into heat. There is, therefore, a limit to the dimensions of the small vortices and to the increase of pressure due to vortex motions, so that the latter increase is quite small.

In this way we see that the local barometric pressure should increase slightly in the early morning hours up to the time when the general turbulence is greatest, which time, as shown by optical phenomena, is about 10 a. m. in the lower strata of air. From this time on the turbulent condition extends upwards to higher layers, and ceases altogether in the lower air by 4 or 5 p. m., by which time the atmosphere has attained its greatest approach to steady convective equilibrium. After this hour the upper regions cease to send cooler masses down to the earth's surface because the latter and the adjacent strata gradually become cooler than they, until finally, by or before sunrise, nearly all convection has ceased, the cold air lies stagnant on the earth's surface; any slowly cooling air left above *BBBB* remains for a while warmer, and the observed barometric pressure is very nearly that due to the simple weight of the superincumbent atmosphere.

6. The diurnal variations of the barometer were analyzed by Lamont into a diurnal and semi-diurnal wave-like oscillation. The former is very variable for different places; the latter is remarkably constant. As to the origin of these oscillations Espy (Am. Phil. Soc., 1817, and Journal Franklin Institute, 1828, new series, Vol. I, p. 278), Davies, Kreil, and Blanford agree that the morning maximum of the semi-diurnal oscillation agrees with the time when the temperature of the air is rising most rapidly.

Blanford (Proc. Royal Soc., London, April 26, 1888) shows that the evening maximum of the semi-diurnal oscillations agrees with the diurnal minimum in the daily variations of cloudiness and rain-fall. In Journal of Asiatic Society of Bengal, 1879, XLVIII, Part II, page 46, he states the striking analogy between the evening storms, commonly known as northwesters at Calcutta, and the thunder-storms of the European summer (and we may equally add, of the United States). Whether these storms are accompanied by rain, hail, or simply dust, they have still the same general characteristics and causes: when the diurnal wind slackens in the afternoon and temporary calm occurs, the storm-cloud is seen in the west or northwest, advancing and apparently growing upwards with great rapidity; new formations continually occur in advance when the nimbus is formed above any locality; the passage of the cloud overhead is speedily followed by violent gusts of wind, raising clouds of dust; the barometer which has previously risen rapidly suddenly falls, the temperature falls, and also the vapor tension.

Blanford, like Mariotte, Espy, Henry, and many others, saw that the gusts blowing out from under the storm-cloud form an eddy or cylindrical vortex, whose axis is horizontal and whose impulse is furnished by the air dragged down by the rain, but not invariably so by rain, since less violent ones occur when no rain is falling, the essential feature being a downfall of heavier air to the ground and its outflow bounding along the earth's surface.

Dr. C. L. Henry (Bibl. Univ., 1860, Vol. IX, pp. 351) assumes that the mass (inertia) and resistance of the ether of space is appreciable, and by resisting the earth in its orbital movement causes an excess of pressure on the front and a deficit in the rear. This accounts only in a general way for the most general feature of the barometric variations, and does not explain any of the local peculiarities of distribution over the earth's surface; it is, moreover, unlikely that the ether has any such inertia as is needed to explain the periodical change of 0.10 inch of pressure.

Kreil finds that the barometric maximum coincides with dispersion of clouds after sunset, and concludes these both are due to the compression that the lower atmospheric strata undergo in consequence of the general cooling, contraction, and subsidence of the atmosphere. This implies that an additional mass of air flows from all sides upon the contracting cool air, which flow would, he thought, increase the pressure, ignoring any effect of cyclonic rotations.

Prof. F. Augustin (Sitzb. Kon. Böhm. Gesell. Wiss., 1881 and 1882) shows that the diurnal or hourly changes of temperature suffice to explain the variations in both rain-fall and pressure at Prague; he concludes that in general the pressure rises when the changes in temperature increase, and fall when the hourly changes of temperature are becoming less, so that the diurnal changes of pressure depend principally on solar heat and terrestrial radiation. Again, he finds the daily maximum and minimum rain-fall agrees also intimately with the daily oscillations of the barometer and the hourly changes of temperature.

Buchan (Royal Soc. Trans., 1880) shows the greater diurnal amplitude over continents and the influence of latitude.

Prof. Dr. Hann (Abh. K. Akad., Vienna, 1889) gives the fullest and best data at present available, and finds evidence of the effect of the direct absorption by the upper atmosphere of solar heat.

Angot (Paris Comptes Rendus, 1888) finds a lunar and solar tidal phenomenon.

7. In regard to the preceding and other explanations of the origin of the diurnal barometric period, I would remark that in general the atmospheric pressure at the surface of the earth can only be affected to an inappreciable extent by the inertia of ordinary, very slowly ascending and descending currents (see, among others, Ferrel Meteorological Researches, Part III, page 8, or Sprung Lehrbuch der Meteorologie), and equally is the effect inappreciable of horizontal rectilinear winds. (See Finger, 1877.) Neither are diurnal variations of temperature or vapor able to produce any but small fleeting effects. The actual weight of vapor daily added to or condensed from the air can produce scarcely an hundredth inch change in the barometer.

The only important pressure changes due to the motion of the air are those produced by curvilinear movements as in vortices and cyclones, or by resisting masses, or by the rotation of the earth, so that any explanation of the diurnal variations of the barometer that is not deduced from hydrodynamic considerations is of questionable value.

The solar and lunar tides in a quiescent atmosphere are barely appreciable (see Chap. I, ¶ 7); the tides or perturbations in a moving fluid, such as the trade winds, have not yet been studied but are probably also inappreciable.

Another clearly hydrodynamic phenomenon has been already considered in the paragraph relating to pressure due to vortex motions.

But besides these we have far more important phenomena based on the general motions of the air as will be explained in the following sections.

8. In a preceding section we have considered the ascending and descending currents due to differences in density, as taking place in an atmosphere that is otherwise at rest and as depending for their origin principally on the temperature. We will now consider a hydrodynamic effect that must be produced by this vertical interchange between layers of air that have different horizontal velocities.

We have here a problem similar to that of the flow of water in rivers; and as there we consider the changes in level, so here we consider the changes in pressure and velocity that are caused by mutual interaction of horizontal currents, and the roughness of the earth's surface, as giving rise to what we have called fluid friction. It is, as we have said before, confusing to speak of the friction of the air on the earth's surface as though there were some similarity between it and the friction of sliding solids; there is little similarity even in the case of air flowing over smooth water and still less in air flowing over the ground; if the surface of the water or ground were absolutely plane we should still have to consider that the air in contact with the surface is firmly held there and that the layers above it slide on each other, thus giving rise to the internal friction or viscosity that we have treated of; but when the surface is undulous or rugged, or when being plane, it has appreciable local variations in its power of resisting the motion of the air, then the new class of resistances arises which we have called friction. As these resistances would still exist in a perfect, namely, a non-viscous fluid, theoretical investigations into ideal fluid motions already begin to throw some light upon them.

9. When the wind strikes an obstacle, as a plate *AB* (Fig. 16), there is at once set up in the neighborhood of the plate a re-arrangement of the velocities of the wind particles; instead of moving with uniform velocities as they do at *LL* and *MM* some will be found moving faster and some slower; if the plate be very small in reference to the section of this stream of gas, then the velocities at *L* and *M* will be nearly the same. On observing the force required to hold the plate in place one would at first conclude that a corresponding amount of energy had been abstracted from the flowing particles of wind, producing a corresponding diminution of the energy of the wind at *M*, but the force required to hold the plate in place, and which we ordinarily call its resistance to the fluid, is not so much due to energy abstracted from the moving particles as it is to a new distribution, transformation, and retransformation of the two forms of energy "potential" and "kinetic" within the flowing mass. Thus on the windward side of the plate at *C* the

motion of the air is checked, kinetic becomes potential energy and static pressure is increased; at *D* motion is increased, potential is transformed to kinetic energy and static pressure is decreased.

If the velocity is great then on the leeward side near *H* there is formed a discontinuous space within which the pressure is very low; similar spaces occur between *a* and *b* as at *c* and *d* with intermediate spaces of higher pressure. The region *ABH* on the leeward side is filled with whirling masses whose total pressure against *AB* is less than what would obtain if the air were quiescent. The fluid between *B* and *K* is moving more rapidly, and between *H* and *I* less rapidly than at any other portion, thus making it possible for the stream to have about the same velocity at *L* and *M*. Therefore the so-called resistance experienced by the plate is principally the difference between the total static pressures on its sides and edges, and the difference in the kinetic energies of *L* and *M* is due to the loss by viscosity and by impact in tumultuous vortex motions. If an anemometer were placed anywhere within the region *ABHI* it would show a diminution of wind velocity precisely similar to what is frequently observed in meteorology among the buildings of a city or among the hills and valleys of the country, and although it is not so common for meteorologists to note the increase of velocity corresponding to the region *BK*, yet such must always exist; in these streams of increased rapidity begin the large class of ascending currents, due to impact, the existence of which will be easily recognized by the observer by watching the motions of smoke and light objects, and by considering the various cases sketched at Fig. 17, 18, and 19.

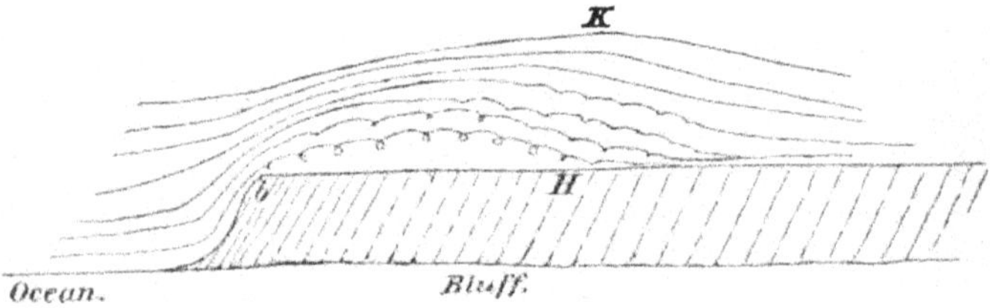

FIG. 18.

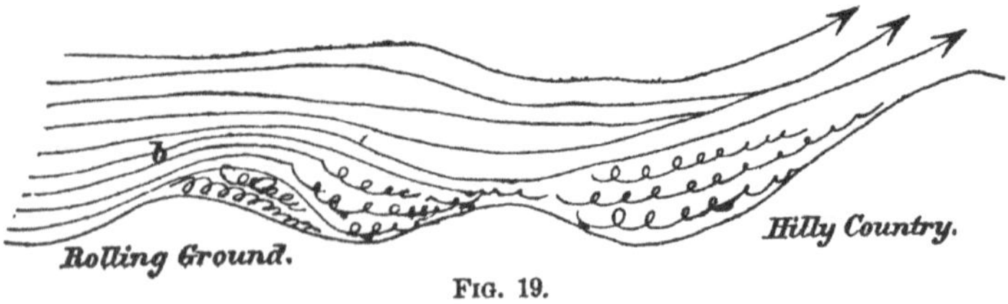

FIG. 19.

10. The effect of every inequality on the earth's surface is to make a region of small wind velocity on the leeward or protected side, and to

divert upward the motion of a portion of the air; this upward dispersion continues so long as the momentum of the moving mass of air holds out. These upward movements correspond closely to those that are also found in agitated flowing water. Thus a stone (*m*, Fig. 20) or other obstacle at the bottom of a river is the cause of currents within the water that manifest themselves in the standing waves *www* and the intermediate troughs (*ttt*, Fig. 20).

11. When the wind blows over the ocean the surface of the water being a flexible boundary between two fluids is in a state of unstable equilibrium so far as the motions on either side of it are concerned, by reason of which the slightest conceivable disturbance produces what are known as capillary ripples (*c c*, Fig. 21) whose dimensions are determined by the tension of the surface layer of water; the wind acting on these produces larger disturbances or the ordinary ripples whose wave lengths do not exceed 1 or 2 inches, while heavier winds produce every variety of wave, surge, swell, breaker, etc. The wave velocity depends, as is well known, principally upon the density and depth of the water and slightly on the wind, but the amplitude of its vertical motion depends upon the density of the water and velocity and density of the wind.

The instability of a thin surface dividing two portions of moving air is beautifully illustrated on the waves which run from one end to the other of a flag (see Fig. 22) freely flung to the breeze. In this case the least disturbance of the smooth flowing wind, such as that introduced by the flag-staff, the inertia of the flag, and the skin friction of the air flowing in contact with the cloth, causes a series of cylindrical vortices in the air which roll along both sides of the flag with their axes nearly parallel to the staff, and the flag which forms the boundary surface separating these vortices is thrown into the beautiful series of moving waves. When the flag is supported on a horizontal rod the weight is added to the forces to be considered and the waves are longer and slower than when the rod is vertical.

Another striking illustration of the ease with which vortex motion is formed in the air may occasionally be seen by observing the smoke from a very tall chimney on a windy day. The illustration (Fig. 23) shows such a series of many vortices observed by me (in August, 1877, in England) existing to the leeward of a factory chimney 300 or 400 feet in height. Evidently the wind blowing swiftly past the chimney formed a slender vortex *bc* whose dimensions were larger at the top where the wind was strongest and diminished downward toward the ground, notwithstanding that the chimney tapered upward very slightly, and which finally ceased to exist at an elevation *c*, where the wind was not strong enough to produce a permanent discontinuous space behind the chimney; and not one only, but fifteen or more well defined similar vortical columns stretching in a series a mile to the leeward of the first, diminishing regularly in length and distinctness to the end. Of course these would have been invisible to the eye had it not been for the slight column of heavy smoke issuing from the chimney, and whose particles, although apparently sucked down into the vortex, were in part at least carried down by their own weight and thus scattered widely over the surrounding country.

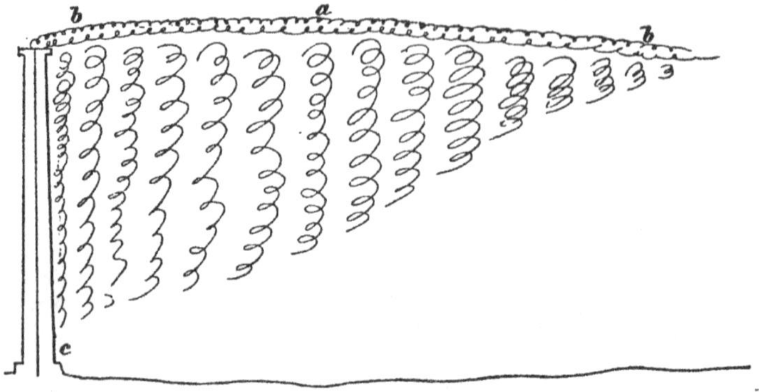

FIG. 23.

12. Returning to the consideration of the air blowing over the waves, we observe that when the wind strikes the windward side of the wave three things occur: First, a portion of the water on the wave surface is pushed forward, thus making the wave a little higher and diminishing the steepness of the sloping surface on the windward side, but increasing the steepness of the slope on the leeward side; second, the wind is forced upward over the crest of the wave, causing the water to comb over, while the air continues on up higher; third, the action of the wind on the windward side causes that to sink all the more rapidly to form the immediately following trough. The vertical amplitude of the wave motion is determined by this reaction of the wind, and by it is also determined the precisely corresponding amount of upward deflection of the air rebounding from the wave (a rebound that is largely utilized by birds skimming the surface of the ocean). Therefore, on the sea as on the land, systematic but smaller upward deflections of the atmosphere exist due to the irregularities of the surface and similar to the small standing waves of running streams. The vertical waves visible in the flag and the vertical vortices to the leeward of the chimney are paralleled by horizontal waves and by vortices whose axes are inclined at all angles and form to the leeward of every irregularity on the surface of the earth.

13. It follows from the preceding that the observer at the surface of the earth is at the bottom of a mass of whirls, whose diameters and positions are perpetually varying, and which derive their force from

the more steadily moving air just above them. The force and direction of the wind that he experiences will vary more and more in proportion as the velocity of rotation of the whirls is greater or less than their velocity of translation. Thus *c*, Fig. 24, is the center of a whirl whose

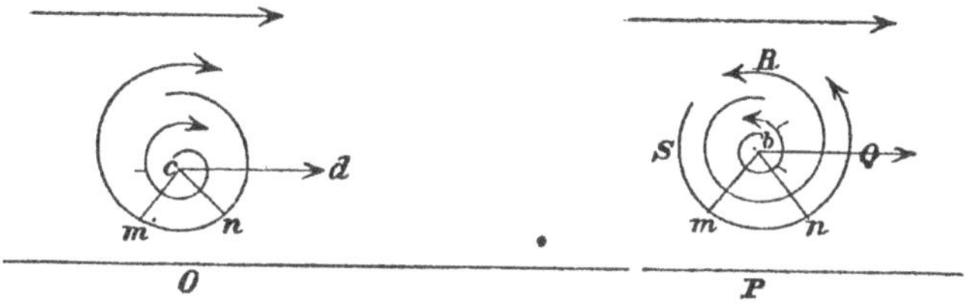

FIG. 24.

axis is nearly horizontal, and for which *cd* is the velocity of translation and *mn* the velocity of rotation. The observer at *O* will experience a velocity $cd-mn$ and in a direction uniform with *cd;* if, however, *b* be a center of a whirl whose axis is nearly vertical, then observers at *PQRS* will experience winds whose velocities are respectively $bd+mn$, $\sqrt{b^2d^2+m^2n^2}$, $bd-mn$, and $\sqrt{b^2d^2+m^2n^2}$, and whose directions will vary between the forward motion at *P* and a possible opposite motion at *n*, thus giving rise to the familiar phenomena of the perpetual variations in force and direction observed during high winds.

When the wind is blowing over the smooth surface of water or prairie land the motions in the air produced by the slight roughness in the surface of the earth are similar to those observed at the surface of the hull of a vessel moving through the water. Between the hull and the quiet water, at a distance from it, is a comparatively thin layer of water that is not quiet, but has a motion which is both rotary and translatory as represented by the minute whirls shown in Fig. 25 clinging to the side of the hull. If the surface of the boat were perfectly smooth the diameters of these whirls would depend on the velocity of the boat and the viscosity of the water; but as it is not perfectly smooth they are somewhat larger, and apt to be present at low velocities where the motion of the waters would otherwise have been in parallel lines such as are consistent with the stable movements explained in a previous section. The velocity of the translation of the centers of the whirls, and in fact of the whole of the thin layer of water involved in them, is about one-half of the actual velocity of the hull, so that when the vessel has moved over the distance *d*, a mass of water equal to that of this layer has been carried a distance $\frac{1}{2}d$, and the energy required to accomplish this must have been furnished by the driving power of the vessel. This is one of the ways water offers a resistance to the passage of the vessel, and as the resistance is evidently proportional to the amount and roughness of the surface involved it has been called skin friction where the use of the word friction implies the enlargement of our definition of that word so as to include any form of resistance to a moving body.

The subject of skin friction has been experimentally investigated by the late William Froude and his nephew R. E. Froude, from whose results the following illustrations are quoted:

Ship.			Wave resistance.	Skin resistance	Total resistance.
No.	Displacement.	Velocity.			
	Tons.	*Knots.*			
1	2,634	13	3.2	5.8	9.00
1	3,634	14	6.15	6.6	12.75
2	3,026	13	2.45	6.95	9.40
2	3,626	14	3.15	8.00	11.15
3	(*)	20	1.1	1.2	2.39

* A torpedo-boat, 125 feet long; displacement small.

In other words, the work done by a vessel in pushing the water aside against gravity, whereby waves are formed, is sometimes of equal or greater importance than the work done in giving a translatory motion to the layer of water that is close to the vessel. Consequently if the vessel were stationary or deeply submerged and formed no waves and the water only were in motion then the resistance experienced by the vessel would represent that energy which is lost by the layer close to the skin of the hull in dropping from its full original velocity down to the much lower average velocity within this layer of vortices.

A similar phenomenon is observable in flowing water behind any obstacle, such as a bridge pier. If behind such obstacle there be a space of comparatively quiet or dead water, while the main stream flows on steadily on either side, there will be observed between the dead water and the flowing stream a region in which little whirls are formed, as at *w*, Fig. 26, the centers of which move down stream with about one-half the velocity of the main current, while the particles of water within each whirl have a uniform rotatory velocity, such that when they are on the mid-stream side they are going downward with the full velocity of the current, but when they are on the dead-water side they have temporarily no translation down stream at all.

The flow of air over smooth land and over the ocean presents such an analogy with that of water that, as a first approximation, we may apply the laws of impact and skin friction to the atmospheric movements in the strata that feel the effect of the earth's surface.

Similarly in the upper strata of air, when, as is frequently the case, we meet with superimposed horizontal currents of air moving in different directions, or when we deal with the upper surface of a lower current, above which the air is comparatively quiet, we have then a very similar case to the preceding, and rolling whirls are formed at the dividing surface between the two layers, which rolls are invisible to us except in so far as they give rise to many varieties of clouds when the conditions are proper as to moisture.

14. Passing from smooth horizontal ocean and prairie to the case when the surface of the earth is quite rough we see that the action of hills, shore-lines, cliffs, etc., on the wind is similar to the action of the large obstacles at the bottom of a stream of shallow running water, which latter has been observed and experimented upon with the following results: The effect of an obstacle at the bottom of a stream is not merely to produce local whirls but to divert the course of an appreciable mass of water from its otherwise horizontal movement, and to send it whirling and slanting upwards and sidewise while other water descends to take its place below; as this mass starts upwards with a horizontal velocity less than that of the water into which it intrudes and with which it mixes it becomes an obstacle to the latter and is pushed along by it so that the diverted uprising current (called a jet in hydrodynamics) has its velocity increased while the upper layers that are pushed aside by it either descend to the bottom or go sidewise to the banks of the river and have their velocities correspondingly diminished. This process goes on till the rising jet is broken into fragments and thoroughly mixed with the mass of the stream, with which its velocity then becomes identical, but which latter is never quite equal to what it would have had as that due to the force of gravity acting upon an ideal smooth stream of the same slope and with parallel filaments.

Although the greater quantity of water thus pushed from the bottom and the sides of a river bed into the central portion attains a uniform velocity when it reaches the body of the stream, yet there is always here and there a more considerable mass, as at *m*, Fig. 27, that will rise to near the very surface of the river before it receives the maximum velocity it is capable of, and will even make a standing wave at *S* or a "rush" at *NR*, and a vortex at *V*. Moreover the slow-moving water at the shallow sides of a stream is continually being thrown in towards the center (the surface slope of the cross section even contributes to the maximum axial velocity, of the stream), the result of which is that the surface of the stream is covered with masses of slow-moving water that are spreading out on all sides and making room for fresh supplies pushed up from the irregular bed below. If, therefore, the down stream velocities be measured at different points in a section across the stream it will be found that the maximum velocity occurs some distance below the upper surface; the minimum velocity is at the bottom and an intermediate velocity prevails at the top. The fact that the surface velocity of a river is less than that a short distance below was formerly explained as due to the drag of the air on the water notwithstanding the fact that it exists even when the wind is blowing down stream much faster than the water is flowing, but the theoretical conclusion that it must really be the effect of the tearing off from the bottom of the stream of those masses of sluggish water that subsequently spread out over the surface is fully confirmed by the study of the flow of water in experimental channels and in pipes where air has no access.

15. Beside the flow of water from smooth pipes and channels where viscosity is introduced, and the parallel case of air steadily ascending in straight lines in minute streams from heated points, and the turbulent flow of air where the mixture of sluggish and rapid movements is initiated by uprising heated currents, we have now to consider a fourth case, viz, the flow of water or air where the mixture of rapid and sluggish currents is produced by impact on obstacles; this has been studied from an experimental point of view by Osborne Reynolds (1884), J. J. Thomson (1876), and others, but from a mathematical point of view by E. B. Hagenbach (now Professor Hagenbach-Bischof of Basle) (in 1860), Boussinesq (1873-'77), Sir William Thomson (1885), and others. The recent studies of the latter on what he happily terms "turbulent motion," and which is a systematic regular form of what Poncelet and Boussinesq call "tumultuous motion," have a direct bearing on our subject.

16. We shall resolve Thomson's turbulent motion into three portions, depending upon the origin and nature of the whirls, namely: First, the very regular whirls close to every smooth surface caused by slip and viscosity and constituting skin friction; second, the whirls permeating the fluid and due to differential densities depending principally on temperature, and which may have their origin anywhere in the liquid, as at the top and bottom of rivers, or bottom and cloud layer of atmosphere; and, third, the whirls produced by impact against irregularities of the surfaces that confine the fluid and the character of which whirls depend on the size and shape of the channel and its obstacles and the velocity of the flow.

"Convective friction" is a general term to include all these three methods of loss of energy due to turbulent flow; it results from the wave motions and the vortex motions attending skin friction, impact, differences of density and lateral currents; it is peculiar to discontinuous and unstable fluid movements, while viscosity, or internal friction, is that due to rectilinear and sliding movements, and is peculiar to the steady and stable motion of fluids; convective friction could exist in a non-viscous or perfect fluid, and is what I have called fluid friction par excellence in my Treatise on Meteorological Instruments. Convective friction depends largely on the existence of vortices, but viscosity is that which tends to prevent vortex and other discontinuous motions, or to destroy them when they exist and to maintain the fluid motion in a state of stability or of parallel laminar flow. The condition of such stable flow, when a fluid meets a fair shaped obstacle wholly immersed in it, or when it flows between two surfaces, must be that the curvature of the outlines of the obstacles or of the surface must conform to the lines of flow of the fluid at the given velocity.

17. Hagenbach, restricting his studies to movements in cylindrical tubes full of water, considers the diminution of energy with which water flows through a tube as the sum of two components, namely, the resistance due to viscosity and the resistance due to the lateral movements by which sluggish moving water at the sides of the tube is transformed to the central portions; this is the process of convection of energy that I have called convective friction.

In Hagenbach's works and the experiments of Poseuille, Hagen, Reynolds, etc., the temperature of the water within the tubes was assumed to be and apparently was kept very uniform, so that the convective friction was due entirely to impact against the slight irregularities of what was meant to be smooth tubes. In the experiments of Hagen, Bruning, Dubuat, Woltman, Michaelis, Darcy, and others, in small artificial channels whose surfaces were roughened at will by the experimenter, and in which the air and sunshine produced differences of temperature, the convective friction due to currents produced by changes of density depending upon temperature was probably appreciable, but was far less than that originating in the impact on the roughened sides. In observations on the flow of water in natural rivers, as made by Brahms, Chezy, Dubuat, Prony, Brunings, Humphreys and Abbot, Buffon, Darcy and Bazin, and many others, the effect of convection due to temperature must have been appreciable, as in fact must generally be the case with observations made in the open air and exposed to the full but variable sunshine.

18. The temperature and therefore the density of a mass of water flowing in a river varies both longitudinally with the stream, as well as vertically and horizontally in every cross section. These variations are due partly to the varied temperatures of the water supplied by the springs, branches, and feeders of the main streams and the melting of ice; partly to the direct accession of rain-fall having various temperatures, and especially to the periodic variations caused by solar and terrestrial radiation and to the non-periodic variations in the temperature of the wind. If the wind is warmer than the water the water temperature at the surface will tend to be warmer than below, and the vertical convection due to temperature and density is diminished or annulled, while that due to impact may still remain; consequently the layer of maximum velocity will be nearer the surface of the water during a warm wind than when a cold or dry wind is blowing, since the latter will increase the convection by its cooling action on the surface. This is true whether the wind blows up stream or down stream, but if the up-stream winds are always warmer and the down-stream winds colder than the water, as in the Mississippi River, then we have a change in the depth of the layer of maximum velocity apparently due to the relative direction and

force of wind and current, such as has been observed in some but not all natural rivers, but which is really due to the relative temperature of the air and water. Our convective friction in a running stream is therefore to be resolved into two parts, *i. e.*, that due to temperature and density, and that due to impact.

19. When, however, we pass from a stream of water to the case of the wind, and consider the horizontal motion of the atmosphere, we see at once that, on account of the proportions of the width, depth, and mass of the stream, the relative amount of turbulent motion is very much larger in the wind than in deep rivers, and, again, that the amount of convective friction in the atmosphere due to impact is of an importance almost equal with that due to temperature and density, and this is true for atmospheric movements over both land and ocean; the latter offers slight changes of temperature to produce convective friction due to density and slight inequalities on its surface to produce convective friction due to impact; the land on the other hand offers large variations of temperature, large orographic irregularities, and correspondingly large convective frictions.

I have attempted to attain some idea of the relative amounts of the resistance due to viscosity, impact, and temperature, from the study of the observations on streams of water discussed by Hagen in his memoir of 1876, "On the uniform movement of water."

The following notation is used by him:

$\alpha =$ the gradient of surface of the water or the amount of fall per unit of length.

$\tau =$ the mean radius of the area of a section of the stream divided by its perimeter.

$C =$ the mean velocity or the discharge in one second divided by the area of the section.

Hagen finds that for a great variety of sections, velocities, and gradients the average velocity of discharge will be represented by the formula:

$$C = k\alpha^x\tau^z$$

where k, x, and z are determined from observations. This formula will numerically represent the observations of Darcy and others in artificial canals of small dimensions if we write it thus:

$$\text{(I)} \qquad C_1 = 4.9\alpha^{0.20}\tau^{1.00}$$

In order to represent the observations made in large rivers and streams Hagen finds that his formula must be written:

$$\text{(II)} \qquad C_2 = 3.34\alpha^{0.20}\tau^{0.50}$$

Finally for the smallest experimental troughs he finds

$$\text{(III)} \qquad C_3 = 3.0\alpha^{0.20}\tau^{0.67}$$

In other words the effect of the passage from the smallest troughs to canals of moderate size has been to introduce the factor $1.5\tau^{.33}$ by which equation III must be multiplied to produce equation I; and, again, the passage from moderate sized canals to natural streams has been to introduce the factor $0.67r^{-0.5}$ by which equation I must be multiplied in order to produce equation II.

The first factor $1.5r^{.33}$ is, I think, due largely to impact on the slightly roughened sides of the canals, and it expresses the principle that the larger the mean radius r for a given constant flow of water, the greater will be the effect of a constant amount of impact on the obstructing obstacles.

The second factor $0.67r^{-0.5}$ required by the passage from artificial canals to natural rivers, represents, I think, the amount of additional disturbance introduced by convections due to density, and expresses the principle that for a constant mass of water and a constant amount of disturbance by the heat of the sun and wind, the effect on the mean velocity will diminish as the square root of the mean radius of the section increases.

The proceding results can be combined with those of others who have tested their formulæ by observations on the flow in small and large rivers, and who all agree that the resistance which we have called convective friction in so far as it depends on impact alone increases with the roughness of the channel.*

We thus entirely dispose of the idea that the atmosphere offers any appreciable direct mechanical resistance to the motion of the smooth upper surface of a stream of water. The same statement applies to ocean currents; these can not be said to be due to the wind so long as the surface of the water is smooth. It is only by differences of density within the water, or by virtue of the impact of the wind against the windward side of the waves, that currents are formed or surface water driven forward forming feeble surface currents in mid-ocean or altering the level of water in harbors.

20. As regards the relative velocities of different portions of a current, some general results may be deduced from the observations on

---

*Among recent memoirs on this subject I quote the following:

(1) Boussinesq in several memoirs between 1868 and 1875, on the influence of friction, centrifugal force, and tumultuous motions on the flow of water, all of which are collected in his great treatise, "Essai sur la théorie des eaux courantes" (Mém. de l'Inst. de France; Sav. Etr.; T. XXIII and T. XXIV; Paris, 1877).

(2) Prof. James Thomson "On the flow of water" (Proc. Royal Soc., London, 1878, XXVIII).

(3) Osborne Reynolds "On the two modes of motion of water" (Proc. Royal Institution, London, 1884, Vol. XI, and Nature, Vol. XXX).

(4) Sir William Thomson "On the stability of fluid motion" (L. E. D., Phil. Mag., (5) 1887, XXIII, XXIV).

(5) Barre de Saint-Venant, in three posthumous memoirs, "Résistance des Fluides," "Perte de Force vive d'un Fluide," "De l'Influence de la force centrifuge dans le mouvement des eaux courantes" (Mémoire de l'Institut, T. LXIV, Paris, 1887).

The memoirs of Boussinesq and Saint-Venant were first accessible to me after writing the preceding pages.

large streams, which show that in the lower half of the current the velocities diminish rapidly as we approach the bottom and sides, but in the upper portion of the stream where the current is more rapid the velocities in successive filaments vary too much to justify a stronger statement than that the maximum velocity is generally below the surface.

If we take the mean velocity of a horizontal layer across the whole width of the stream Hagen finds that such mean velocities may be represented by the parabolic formula

(IV) $$y = k^{0.5} a^{0.25} x^{0.5}$$

where y = the mean velocity of the whole of any horizontal layer.
x = the distance of this layer upwards from the bottom.
a = as before the gradient at the surface layer.
k = the arbitrary constant.

For very shallow streams over smooth surfaces and for low velocities where convective friction is inappreciable and viscosity only remains, this parabola becomes a straight line; that is to say, for such a case the co-efficients of y and x are each unity.

Bruning's measures of velocities at different depths in large streams gives relations that Hagen has expressed, as follows:

z = mean horizontal velocity of the whole of any vertical column, so located in the stream as to be away from the shallow, sluggish water near the shores.
C = velocity at the surface of the stream at the top of any such column.
τ = the whole depth of water of the column.
t = the depth below the surface of the layer whose velocity is C, or the average of the whole column.

If, then, we use the meter and second as units, we have for natural streams of water:

(V) $$z = C\,(1 - 0.0582\,\tau^{0.5})$$

(VI) $$t = \tfrac{1}{2}\,\tau\,(1 - 0.0492\,\tau^{0.5})$$

In other words $(\frac{z}{C} - 1)$ and $(\frac{t}{\tau} - 0.5)$ are the ordinates of parabolic curves whose abscissas are τ and whose parameters are 0.0582 and 0.0246 respectively.

The observations represented by this formula covered a range from 4 to 22 meters in depth, and from 20 to 70 meters per second in velocity; therefore, in applying them to the air we may expect them to hold good for a similar range of velocities, but for depths eight hundred times greater than that of water, which abundantly covers ordinary cases of meteorology, excepting only, that in the latter case the convections due to temperature and impact are relatively much larger than in water.

In atmospheric movements the velocity of the horizontal current at the level of the tops of the cumuli or above them represents that at the surface of the river. The widths of the currents of air are usually so great in comparison with the depths that the condition of being distant from the shores of the river is abundantly fulfilled. These formulæ, therefore, when the constants have been properly determined, should enable us to determine the average and the total horizontal flow of air in a current whose velocity is known at a considerable height above ground such as is given by the velocity of motions of the clouds.

21. In seeking to apply these results to the atmosphere the first point of difference is, that in it we have nothing precisely corresponding to the upper boundary surface of a stream of water; only in the case of the underflow of dense air, as experienced in our northerly winds and especially in our cold waves, do we have a tolerably well-defined upper surface separating the dense air from that which is above it and considerably lighter. When convection currents exist in the water they can not rise much above its upper surface (*i. e.*, they make only slight waves); when such currents exist in the air it is by no means impossible for them to continue upward indefinitely into regions where the atmosphere would have been too rare to support them had they not become specifically lighter by reason of the evolution of the latent heat of condensation. When such an extreme ascent occurs, however, the upper portion of the rising column eventually reaches a region where its buoyancy is less than that of the surrounding air, as in the case of the tops of the cumulus clouds, which then flow to one side very much as when streams of water rise (or "rush") up from below and spread out on the surface of the river, forming the phenomena known as "rushes" among river men and millers.

In general the ascending columns within air flowing horizontally do not reach the definite upper surface of their own layer, therefore, small cumuli generally show pointed summits as though still penetrating upwards; in this case, however, any cirri that may be present show us the upper limit of the current within which the cumuli are moving and rising; cirri are often formed before the cumuli below them, but in a stratum having its own independent system of convective currents.

Doubtless also in the natural streams of water the greater part of the columns that start upward are broken up before they reach the surface; therefore, in natural streams of air and water we should expect that similar laws would hold good as to relative velocity of layers, and that Hagen's parabolic curve of increase with altitude may apply at least to general averages.

22. Of the few experimental investigations that have been made into the relative movements of the air at different heights above the earth's surface, the most interesting one is that of Archibald, who, by means of anemometers, carried up on kites, has given us valuable data as to relative velocities at elevations between 90 and 1,800 feet above the ground at his place of observation, in Kent County, England, and which was itself 500 feet above sea-level.

The following table is compiled from Archibald's latest communications on this subject (Nature, 1886, Vol. XXXIII, p. 593) and presents the results of forty-two pairs of velocities as measured simultaneously by two anemometers whose vertical distance apart was from 100 to 300 feet. These forty-two pairs have been arranged in six groups, according to the altitude of the lower anemometer. The table gives the mean heights of the anemometers for each group and also the corresponding means of the measured velocities.

The excess of the velocity at the upper anemometer over that of the lower, divided by their difference in altitude gives the rate of diminution of velocity per foot of elevation; these ratios, as given in column 9, show that the rate diminishes first rapidly and then slowly as we ascend.

Group.	No. of observations.	Mean height above ground			Mean velocity per minute.			$\frac{V-v}{H-h}$ or $\frac{dv}{dh}$	*n*.
		Upper anemometer. *H*.	Lower anemometer. *h*.	Mean of both.	Upper anemometer. *V*.	Lower anemometer. *v*.	Mean of both.		
		Feet.	*Feet.*	*Feet.*	*Feet.*	*Feet.*	*Feet.*		
1	7	250	102	176	1,617	1,174	1,395	3.0	0.372
2	3	322	128	225	2,232	1,679	1,955	2.8	0.307
3	8	407	179	293	1,705	1,385	1,515	1.4	0.275
4	5	549	252	400	2,107	1,773	1,949	1.2	0.237
5	9	795	481	638	2,192	1,957	2,074	0.4	0.250
6	10	1,095	767	931	2,236	2,096	2,166	0.45	0.194

The ratios in column 9 represent approximately the tangent of the inclination of a parabolic or logarithmic curve similar to that obtained by Hagen for water in natural streams. If on the other hand we compare the actual velocities in their respective pairs we have

$$\text{(VII)} \qquad \frac{V}{v} = \left(\frac{H}{h}\right)^n$$

and the values of the exponent n for each stratum, as given in column 10, show a rapid diminution of the exponent as we ascend above the ground; comparing columns 9 and 10 we find a decided increase of n with altitude as compared with the increase of actual velocities, showing that greater convections occur at higher velocities.

23. For still higher atmospheric currents, Professor Archibald adduces the results of observations of velocities of clouds by Dr. Vettin, of Berlin, to show that the exponent n reaches a limiting value of 0.25 which represents Vettin's velocities between the altitudes 1,600 and 23,000 feet. Except for the general increase of velocity with heights due to the rotation of the earth as deduced by Ferrel (Meteor. Researches, Part 1, Table XI), and which is independent of friction, there is no reason why the preceding exponents should depend upon elevation of the ground above sea-level, provided the ground presents a sufficiently extensive and nearly horizontal surface; only in case the ground is a bluff or table-land, such that the winds bound up over it, leaving

the lower layers entirely unaffected, we must count the elevations from the lower surface of the moving air. The relation above deduced between successive velocities evidently depends on the irregularity of the country, which we will call a local factor, and on the convection due to heat which has a diurnal period of a very general nature, as also a local constant term: this diurnal periodicity is evidently very well marked since almost everywhere in the equatorial and temperate zones the velocity at the surface of the earth diminishes to nearly zero sometime during the night.

24. Archibald finds in each of his six groups, and, therefore, at all heights up to 1,800 feet, a minimum co-efficient at about 2 p. m., and a maximum at the time of his earliest and latest observations, which were made about 8 a. m. and 8 p. m. This, therefore, accords with mountain observations made at still greater heights, which show that whereas at sea-level the wind velocity is a maximum during midday and a minimum during midnight, the reverse is the case at an elevation of 4,000 feet; this latter represents the elevation to which, by convection from below, sluggish ascending columns have risen and by their inertia produced an appreciable amount of drag or convective friction. We shall have to ascend much higher to get above such columns and into a region where the horizontal motion is independent of such sluggish movements below.

25. In the ordinary formula for the flow of a fluid the gradient of pressure is a continuously acting force, and the resulting velocity of the particles of fluid is at first an accelerated motion, but they are speedily brought by friction into a uniform state of motion, so that the gradient of pressure serves as the power which is just able to overcome the various resistances and maintain the uniform flow. The diurnal variation in the temperature, and the corresponding variation in the "thermal convective friction," is that which causes the diurnal variation in the wind at the surface of the earth, causing the lowest wind to be the swiftest when the convection is greatest, and allowing the air at the earth's surface to remain perfectly calm when there is no such convection, therefore, our measured surface wind velocity is not at all that due to the simple mechanical effect of the observed surface barometric gradient, but for a constant gradient the wind velocity is greater during the day and less during the night. It might seem strange that velocity should at any time be greater than that due to the gradient, but this will be seen to be true when we consider that the observed gradient at the earth's surface during steady straight winds represents the gradient prevailing a short distance above the surface in layers which if they move more rapidly than these below must, by convection, impart to the latter a velocity they would not otherwise have.

26. The diurnal variation in Archibald's exponent n, or in the ratio $\frac{V-v}{H-h}$ for any altitude, shows that the difference of velocity between the strata has a diurnal variation, which is undoubtedly entirely de-

pendent on the relative buoyancy, which latter is determined by the insolation. The effect is greatest in the lowest layers when the daily mixture of currents begins, namely, in the early morning, and must become zero during those hours when the vertical temperature gradient is consistent with static equilibrium, but it is also permanently zero at and above the upper limit to which convection currents ascend, and this, as shown by the clouds, is frequently at least ten times the height attained in the measurements by Archibald; therefore, his results can only be regarded as justifying Hagen's formula for the lower layers of a stream. For the upper layers we must refer to direct observation of cloud-motions, and on this point the observations of Vettin and Ekholm and Haegstrom accord with Archibald, all of whom find that the ratio "velocity of the wind at the height of the cloud, divided by the velocity at the surface of the earth," has a minimum at midday and maximum during the night. The results of the observations by Ekholm and Haegstrom during June to August, 1884, give the following general ratios of simultaneous observations:

$$\text{Midday}\left\{\begin{array}{l} \dfrac{\text{Velocity of upper clouds at altitude 5,900 meters}}{\text{Velocity of wind at the surface}} = 2.76 \\[2ex] \dfrac{\text{Velocity of lower clouds at altitude 1,558 meters}}{\text{Velocity of the wind at the surface}} = 1.40 \\[2ex] \dfrac{\text{Velocity of upper clouds at altitude 5,900 meters}}{\text{Velocity of lower clouds at altitude 1,558 meters}} = 2.97 \end{array}\right.$$

$$\text{5.30 p. m.}\left\{\begin{array}{l} \dfrac{\text{Velocity of upper clouds at altitude 6,320 meters}}{\text{Velocity of the wind at the surface}} = 2.90 \\[2ex] \dfrac{\text{Velocity of lower clouds at altitude 1,407 meters}}{\text{Velocity of wind at the surface}} = 1.59 \\[2ex] \dfrac{\text{Velocity of upper clouds at altitude 6,320 meters}}{\text{Velocity of lower clouds at altitude 1,407 meters}} = 1.87 \end{array}\right.$$

These ratios show that from midday to evening the cumuli increase their velocities as relative to the cirri in the ratio of 2.97 to 1.87. This latter conclusion may indicate either one of two inferences: Either, first, the observed cirri were on the upper surface of the same current in which the cumuli were immersed, which current had the greatest velocity at the level between the cumuli and the cirri, as in most streams of water; or, second, the cumuli represented the top of the given current while the cirri represented convection clouds in an independent current above the lower one. The latter is more generally the case, and the velocity of such upper current is therefore dependent on gradients at high levels, as developed by Ferrel in his researches, as before quoted, and as computed by Supan and others.

We conclude, therefore, that the horizontal flow of air for layers below the tops of the cumuli is very nearly as expressed by the exponential forms of Hagen's formula, given in equations I to VI of paragraphs 19 and 20.

27. As the gradient of the median layer in a river determines the force which moves the whole mass of water against all resistances, so the gradient at the median level (namely, that for which the values of z and t obtain, as given in V and VI) in a current of air determines the mean velocity. Such a gradient at a considerable elevation above the earth's surface differs somewhat from that at the surface, as given by our barometric observations, and the ordinary comparison of wind velocity with sea-level gradients, without allowing for resistances, may lead to some confusion. Let the accompanying diagram be the ideal sec-

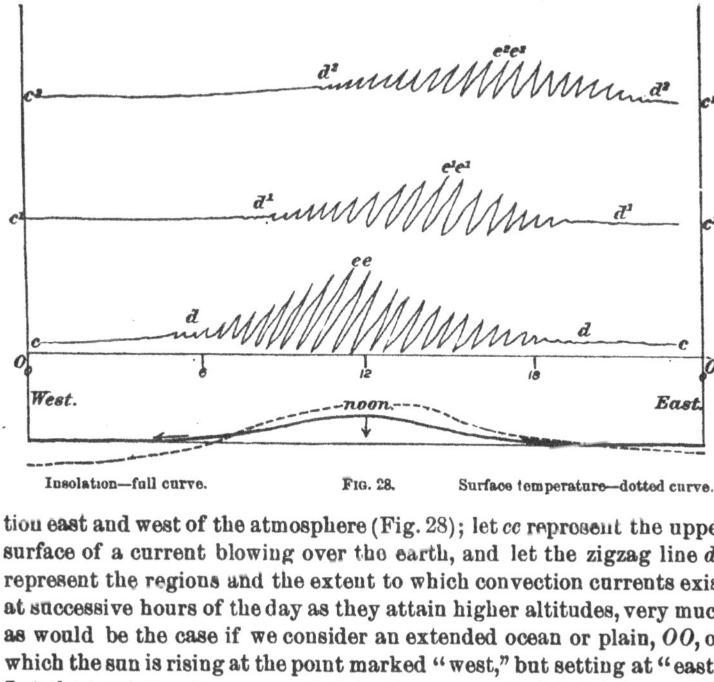

Insolation—full curve. FIG. 28. Surface temperature—dotted curve.

tion east and west of the atmosphere (Fig. 28); let *cc* represent the upper surface of a current blowing over the earth, and let the zigzag line *dd* represent the regions and the extent to which convection currents exist at successive hours of the day as they attain higher altitudes, very much as would be the case if we consider an extended ocean or plain, *OO*, on which the sun is rising at the point marked "west," but setting at "east." Let the insolation be represented by the arrows depicted below. It is now evident that the west wind *cc*, which on the west side of the ocean extends uniformly down from *c* to *O*, will, as it advances, have to overcome the convection resistance indicated by the zigzag lines, and which is a maximum near noon time. The flow of air is, therefore, delayed and as in all other cases of fluid motion the pressure within the stream must rise, the loss of velocity being proportioned to the square root of the resistance, namely, nearly as the square root of the mass of air carried up by convection.

If two masses m_1 m_2 have velocities v_1 and v_2 the velocity (V) of the mixture will be given by

$$\text{(VIII)} \qquad V^2 = \frac{m_1 v_1^2 + m_2 v_2^2}{m_1 + m_2}$$

This resulting V is nearly as the square root of the height of the convective currents, and this is as the square root of τ in Hagen's formula I, paragraph 19, above, if we consider these currents as conical-shaped masses of ascending and descending air. So long, therefore, as the convection is increasing and the horizontal movements of the upper layers is slowing down at elevations increasing up to 10 or 12 a. m. the pressure within the stream must rise, and the upper surface of the air cc will rise slightly to correspond, as when any obstacle interrupts its flow.

We have here another source of diurnal barometric variation and one that like the one above described depends on the action of convection currents, but the present one produces its diurnal *maximum* rise somewhat later in the day than the former, *i. e.*, when its convection currents have permeated the whole lower atmosphere.

28. Still other dynamic periodicities exist in meteorology depending on the general motions of the atmosphere, and on the effects of insolation upon the clouds. As to the former it is sufficient for the present to refer to the differential formulæ for the movement of the atmosphere over the earth's surface as a whole as given in equation 13, page 188, of Ferrel's Recent Advances. In these* the variation of temperature with latitude is expressed as the sum of a series of terms of the form

$$S\, A_s \cos s\theta$$

so long as A_s varies only with latitude these equations give by integration the distribution of wind and pressure over the surface of the earth as dependent on latitude and altitude, and Ferrel has resolved them in a general way for the cases of maximum, minimum, and mean annual temperature, namely, for the approximately permanent or steady conditions of temperature, wind, and pressure that prevail in January and July and for the ideal secular average.

If, however, we consider the diurnal variations of temperature on each parallel of the earth's surface, namely, the diurnal variation of the co-efficient A_s, we see at once that the integration of the general equations would introduce into the values of the pressure P terms of the periodic form, *sine nt* and *cosine nt*. In the case of no friction or resistance these variations will become zero at the poles and a maximum at the equator, but under actually existing circumstances their co-efficients are large for the continents and smaller for the oceans; large in summer

*These remarks apply equally to the more perfect solutions of the general equations as given by Oberbeck in Sitzbericht, K. P., Akad, Wiss., Berlin 1888, page 383.

and small in winter; large in temperate and tropic but small in polar regions; large in clear and small in cloudy weather.

29. The corresponding equations for motion and pressure in a large cyclone (see equation 4, page 237, Ferrel's Recent Advances) similarly express the variations of pressure and cyclonic winds depending on the variations in buoyancy or aspiration towards the central region of disturbance, and, as the buoyancy has its diurnal and orographic variations, these also introduce diurnal and local barometric terms, which, however, are less important than the preceding in proportion as the diurnal variation in storm development or buoyancy of storm cloud due to insolation of the clouds is less important than the convections due to insolation of the earth.

30. The above four dynamical causes for the diurnal variation of the barometer, other than that depending on variable insolation, are, as I have long since stated, those that seem to me to offer sufficient explanation of the existence and permanent peculiarities of the diurnal barometric fluctuations. I should briefly designate them as: first, vortex terms due to the presence of small whirls in the atmosphere; second, convective resistance terms due to the influence that sluggish rising matters have in diminishing the velocity due to gradient and converting kinetic energy into potential energy; third, general insolation terms or the effect of the variable diurnal insolation upon the general atmospheric buoyancy and general movements; fourth, cyclonic insolation, or the effect of local and diurnal changes of movements within cyclones.

31. In the preceding general hydraulic law for the horizontal flow of water and air the pressure gradient (α) refers to the rate of change in pressure in the direction of the movement of the air; this is a differential pressure due to gravity or weight and is supposed to be the only force that produces the motion; it is, therefore, distinct from the gradient (G), ordinarily spoken of in meteorology, which is called the barometric gradient and which is measured normal to the isobars, and is the result not the cause of the motion of the air. If the wind is inclined to the isobar at the angle i the relation between α and G is given by modifying Ferrel's formula (Recent Advances, page 282) which involves the latitude, velocity, friction, and inclination.

32. Our hydraulic equation finds its simplest application in the flow of cold air at the surface of the ground, as exemplified in the phenomena known as "cold waves" in the United States, and from observations on these alone we may make a first approximation to the values of the constants in that equation.

But the dependence of the constants on insolation, orography, and wind direction, is best brought out by the comparison of such winter observations with similar movements in summer time.

33. Before such numerical comparisons are made let us consider, from a deductive point of view, the meaning of the terms that enter our

formula and the simplest method of deducting the co-efficients. If during a steady wind the barometric pressure in the line of wind motion varies on the average throughout the whole mass of moving air at the rate of α_g barometric units per degree of the great circle, then the motion of the wind corresponds to a real moving force that is equal to this linear gradient diminished by α_r, or the linear gradient that corresponds to the unknown resisting influences that are hindering the movement of the wind. Thus the constant moving force is $\alpha_g - \alpha_r$, consequently the resultant uniform velocity as an average for the whole mass is given by the expression

$$V^2 = 2g\,(\alpha_g - \alpha_r)$$

or

$$\text{(IX)} \qquad V = \sqrt{2g\alpha_g}\left(1 - \frac{\alpha_r}{\alpha_g}\right)^{0.5}$$

Now the resistances themselves, or the term $\frac{\alpha_r}{\alpha_g}$, depend upon the actual velocity of the wind, and, as a rough approximation, we may assume them to depend upon the square of the velocity, and adopt the following notation and minor assumptions:

Let $\mu =$ the co-efficient of viscosity of the air for the prevailing average temperature, this affects the whole air throughout the depth τ.

Let $\iota =$ the co-efficient for impact depending on the orographic irregularities or the roughness of the ground in the line of motion of the wind.

Approximately assume $\iota = xy$, where x is the ratio of the average height of the obstacles to the depth of the fluid, and y is the *sine* of the average angle between the normal to the obstructing surfaces and the direction of the motion at the time of impact. This resistance also affects the whole depth of the fluid.

Let $\delta =$ the co-efficient of upward convective movements assumed to extend upwards to the top of the stream of air and depending on the temperature of the air at the ground and on θ, or the existing rate of diminution of temperature for a unit increase of altitude.

Sir William Thomson and numerous succeeding writers have shown that the lower part of the atmosphere is in indifferent static equilibrium when the rate of temperature diminution (θ) is 0.0099 C. for 1 meter of of ascent, and that in proportion as the rate is greater, or as θ is numerically greater than 0.0099, the conditions are favorable to convection currents, therefore, δ may be considered approximately proportional to $0^\circ.0099 - \theta$.

Let A B C be three co-efficients, to be determined by observations, and τ as before the depth of the flowing air. The expression for the resisting forces may now be written

(X)
$$\frac{\alpha_r}{\alpha_g}=(A\mu+B\iota+C\delta)\,\tau^{-1}\,v^2$$

Substitute this in the preceding value of the velocity V, and we have as a rough approximation

$$V=\sqrt{2g\alpha_g}\left[1-\tfrac{1}{2}\,(2g\alpha_g)^{-\frac{1}{2}}\,(A\mu+B\iota+C\delta)^{-\frac{1}{2}}\,\tau^{+0.5}\right]$$

or by introducing two generalizing factors m and n.

(XI)
$$V=m\sqrt{2g\alpha_g}\left[1-n\,(1+A\mu+B\iota+C\delta)\,\tau^{0.5}\right]$$

Comparing this with Hagen's formula (V, paragraph 20),

$$Z=C\,(1-0.0582\tau^{0.5})$$

we see that

(XII)
$$C=m\sqrt{2g\alpha_g}$$

and that

(XIII)
$$0.0582=n\,(1+A\mu+B\iota+C\delta)$$

This last expression which holds good for Brüning's observations on flowing water also represents approximately the case of the atmosphere when δ is zero and ι quite small; the determination of the exact values of these separate co-efficients for the atmosphere must be made by intercomparison in a large number of observations of wind and cloud movements, but the general co-efficients as above given from Hagen will serve to show the probability that the total effect of the impact and viscous resistances (or of the co-efficients ι and μ) is to diminish the horizontal movement by a quantity not to exceed one tenth of its full value; the co-efficient δ for density is very variable and in special cases may have a larger effect than both the others combined.

CHAPTER IV.

ASCENDING MOTIONS IN THE LOWER STRATA AS DUE TO BUOYANCY.

1. As before stated the difference between the velocity of the air at the ground and at higher elevations has its origin not only in impact against the terrestrial irregularities, but also in the action of heat and moisture, both of which cause the differences of density that by the action of gravity cause differences of buoyancy, thereby bringing about interchanges between the upper and lower strata, such as disappear only when the vertical and horizontal distribution of density are such as will constitute indifferent or stable equilibrium.

By reason of such vertical interchanges the slow moving masses of air near the earth's surface change places with the rapidly moving air above whereby the slower horizontal motions below come to be accelerated and the rapid motions above are diminished, or *vice versa* if it should occasionally happen that the upper currents do not move so rapidly as the lower.

In this way an interchange of momentum is brought about between the contiguous superposed currents of air; an interchange which is not due to viscosity nor yet to the whirls that characterize fluid friction, nor is it quite similar to the so-called skin friction. In so far as we meet with it in the atmosphere, this interchange accelerates the horizontal movements of the lower strata and retards those of the upper, during the day time, causing the diurnal variation in the direction, velocity, and force of the wind; on the other hand, by reason of the absence of rising buoyant currents during the night time, the upper currents are apparently accelerated, *i. e.*, not retarded, at night. Therefore, by this convective interchange of energy both upper and lower currents come to have diurnal periods that are complementary to each other. This effect is very prominent in atmospheric motions; it increases the movement of air during the day time at the earth's surface and it diminishes the horizontal movement during the day time at points between 3,000 and 15,000 feet above the earth's surface, and I have sometimes called it "convective resistance due to buoyancy" as distinguished from "convective resistance due to currents produced by impact;" in this use of the word "resistance" it may be either positive or negative and the same generalization would have been required had I used the term "convective acceleration."

2. The study of vertical motions, therefore, adds some new features to what has been said in Chapter II, and meteorology has now to

deal with the three following forms of resistance to the horizontal flow of air to all of which the term friction has hitherto generally applied:

(*A*) Sliding friction, otherwise known as slipping friction, adhesive friction, surface hesion and adhesion, or the actual sliding of air along but in contact with smooth solid or liquid surfaces.

(*B*) Internal friction, otherwise known as internal fluid friction, or viscosity, or the resistance that one layer of particles experiences when it glides at a uniform rate past another layer of the same fluid.

(*C*) Convective friction, otherwise known as fluid friction, eddy friction, external fluid friction, vortex resistance, or discontinuous resistance, all of which terms may be analyzed as follows:

(*a*) Skin friction or the energy consumed in communicating vortex motion to a stratum near and parallel to the surface along which the fluid moves.

(*b*) Impact friction, or the energy consumed by impact of the fluid against obstacles moving across its general path.

(*c*) Convective friction, or the convective reactions partly resistances, partly accelerations due to interchange of momentum between small masses of the fluid itself moving in various directions and with different velocities. The various motions of these small masses have their origin in three ways: (1) in deflections by impact or pressure on resisting solids; (2) in static instability due to changes of density and buoyancy caused by changes of temperature; (3) similar instability due to changes in hygrometrical conditions.

3. As regards No. (*A*) we may assume that there is no slipping of the air over continents or oceans, but that the adherence of air to terrestrial objects is very tenacious so that the next layers of air particles must slide over those that adhere to the solid surfaces; therefore, the co-efficient of slip is 1.

As regards No. (*B*) the air is so nearly a perfect fluid that its defect in this respect is a very small disturbing feature in the general motion of the mass; the adhesion of layers of gases to smooth or even polished surfaces, the so-called surface adhesion, surrounds most bodies with a layer of air over which the next adjacent layers may glide so that ordinary attempts to determine the co-efficient of slip or sliding friction of air on solids fail and we obtain only the co efficient of air sliding on air or viscosity. Thus Maxwell, O. E. Meyer, and Obermayer, have arrived at the following numerical results or viscosity of dry air at temperature 0° C:

$$\mu_0 = \begin{cases} 0.0001878, \text{ Maxwell.} \\ 0.0001727, \text{ Meyer.} \\ 0.0001705, \text{ Obermayer.} \end{cases}$$

The co-efficient for moist air is not appreciably different from that for dry air at ordinary temperatures and pressures.

These numbers indicate that if a surface of 1 square centimeter of air particles glides upon another air surface 1 centimeter distant from it at the rate of 1 centimeter a second it will require a uniform pull or pressure or shearing stress of about 0.00018 dynes to maintain a uniform sliding movement. This constant holds good for 0° C. and increases with temperature but is independent of the barometric pressure or the density of the air. Its dependence upon temperature (θ) has been determined by several investigators, as follows:

Maxwell (1860) $\mu=\mu_0 (1+0.00366\times\theta)^{1.00}$.
Holman (1878) $\mu=\mu_0 (1+0.00366\times\theta)^{0.77}$.
Barus (1889) $\mu=\mu_0 (1+0.00366\times\theta)^{0.67}$.

The above figures are expressed in the C. G. S. units ordinarily adopted in the physical laboratories but become more intelligible to the meteorologist when expressed in terms of the barometric gradient. Let the unit gradient be a difference of pressure corresponding to 1 millimeter of the barometer for 1° of the great circle; the latter arc is equal to one-ninetieth of 10,000,000 meters or 11,111,111 centimeters; 1 millimeter of barometric pressure is equivalent to a pressure of 1014000÷760=1321 dynes per square centimeter; therefore, our unit gradient of barometric height is equivalent to a gradient of pressure of 1321÷11,111,111=0.00012 dynes per linear centimeter and this, so far as viscosity at 0° C. is concerned, will maintain a uniform difference in velocity of 0.0001878÷0.00012=1.56 centimeters per second between the bottom and the top of a stratum 1 centimeter thick; this is equivalent to 0.156 meters per second or 0.0347 English miles per hour.

This resistance is inversely proportional to the assumed distance of the layers and directly proportional to the difference of the velocities, so that the above figures may be expressed in English units thus: A wind of 0.0347 miles per hour in a plane 1 foot above the surface of another plane, where the velocity is zero, implies a viscous resistance such as can be overcome by a barometric gradient of 0.00129 inches of the mercurial barometer per degree of the great circle.

4. In order to apply this result to atmospheric movements it is necessary to know the relative velocity of layers of air a foot apart, or the actual phenomena of the atmosphere. In general it has not been possible to determine any difference in relative velocity of two strata a few feet above each other and near the surface of the ground on account of the uncertain errors of our anemometers and the great irregularity of the wind. But the general result of the observations by Stevenson and Archibald, as quoted in Chapter III, shows that for lower strata the following formula will hold good in Great Britain:

$$\text{In feet per second, } v=28.1+0.2\sqrt{\text{height in feet}}$$

$$\text{In miles per hour, } v=18.7+0.13\sqrt{\text{height in feet.}}$$

Computed by this formula we get the figures in the following table, in which the last column shows the differential motions between successive layers a foot apart as we rise in the atmosphere up to a height of 300 feet:

h	v (hourly).	$\frac{dv}{dh}$	$\frac{dv}{dt}\cdot\frac{1}{v}$
Feet.	*Miles.*		
0	18.7	∞	∞
1	18.8	0.066	0.0036
49	19.6	0.019	0.0010
100	20.0	0.007	0.0003
144	20.3	0.006	0.0003
225	20.8	0.004	0.0002
289	20.9	0.004	0.0002

For higher altitudes we have recourse to the direct observations by Archibald, which gave him the results given in the following table, where again the last column gives us the differential movements desired for layers one foot apart:

Average h	v (hourly).	$\frac{dv}{dh}$	$\frac{dv}{dt}\cdot\frac{1}{v}$
Feet.	*Miles.*		
171	16.0	0.034	0.0021
225	22.0	0.031	0.0014
293	17.5	0.016	0.0009
400	22.0	0.013	0.0006
638	23.5	0.008	0.0003
931	24.6	0.005	0.0002

We may, therefore, infer that in so far as the air strata glide horizontally over each other, the amount of such gliding is such as to give differential velocities of from 0.004 to 0.050, or on the average not more than 0.015 miles per hour for layers of air a foot apart at altitudes up to 500 feet, and for general velocities of the wind of not exceeding 25 miles per hour. This differential movement corresponds, according to our previous computations, to a viscous resistance, such as can be overcome by a barometric gradient of about 0.0006 of an inch of barometric pressure per degree of a great circle. On the other hand, the sea-level gradient that actually accompanies a wind of 20 miles per hour, even in the strong winds of Great Britain and the North Sea, is about 2.5 millimeters or 0.1 inch per degree, or 170 times as large as that required to overcome viscous resistances. Therefore, we conclude that the barometric gradients actually observed in the atmosphere are so much greater than is necessary to overcome viscosity that the latter ceases to be of importance in meteorology where the air is free to take upon itself any one of the many forms of discontinuous motion. The true office of viscosity seems to be to furnish an initiating or determining cause of vortex motion, or again, to annul such vortex motions when they take place on a minute scale in the atmosphere, as in fact is always occurring, as explained in Chapter II.

5. In atmospheric motions on a larger scale, such as the dimensions of a square yard or more, viscosity has but little effect on the slower vortex motions; in this case it is the surrounding or boundary circumstances that initiate the circular or vortex motions, in which case, as in the minuter motion, the direction of the rotation may be either positive or negative. When the motions take place on a much larger scale, such as a square mile or more, viscosity is entirely negligible and the direction of the rotations is largely controlled by the rotation of the earth on its axis; occasionally the conflict of two currents or the location of a mountain range or other obstacle may be such as to annul the effect of the earth's rotation, but such local peculiarities are rare and are able to maintain the unequal contest but a short time. When the axis of rotation of a vortex is not horizontal the influence of the earth's rotation is always in the direction of giving the vortex a definite direction of rotation, *i. e.*, one which in the northern hemisphere is opposite to the direction of motion of the hands of a watch or the same as that which prevails in all large northern cyclones, but which in the southern hemisphere is in the same direction as the motion of the hands of a watch or the same as observed in all large cyclones in the southern hemisphere.

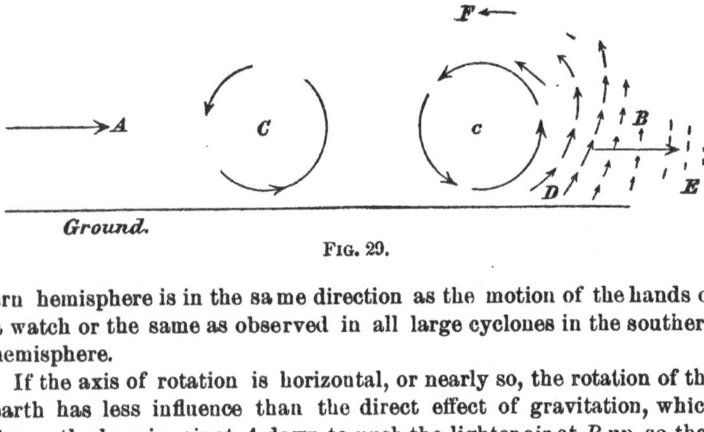

FIG. 29.

If the axis of rotation is horizontal, or nearly so, the rotation of the earth has less influence than the direct effect of gravitation, which draws the heavier air at *A* down to push the lighter air at *B* up, so that the direction of rotation of a horizontal vortex cylinder, such as we sometimes find in the advancing front of a thunder-storm or of a cold wave, is such that the lower portion of the revolving mass is moving forward, namely, in the direction of translation of the whole, while the upper portion is moving backward. This is shown in Fig. 29, where *C c* represent the centres of rotating cylindrical cores whose axes may be 10 or 100 miles long. In such advancing rolls the pressure at *D* in the still air first increases slightly; then the limpid air, having no other outlet, is pushed up before that at *E* feels any pressure; *D* rises and comes to rest on top at *F*, while the rolling heavy core *C* advances and settles as cool air quietly on the ground.

6. A very narrow cylindrical column, such as a water spout or the funnel of a tornado, has some analogies in its motions to the movements of fluids flowing through long cylindrical tubes; the mechanical phenomena in the latter case have been investigated mathematically by Hagemann, Bischof, Stefan, Boussinesq, Sir William Thomson, and others, but the experimental investigation by Osborne Reynolds has been particularly instructive.

When water is maintained in a state of steady flow through a smooth cylindrical tube, the nature of the tube and liquid being such that the layer adjacent to the tube adheres to it while those in the interior slide over their neighbors, then the viscous resistance thus introduced causes the velocity to decrease steadily from the centre to the sides. Up to a certain limiting velocity this condition of motion is stable—any slight disturbance in the earlier portion of the flow through the tube diminishes as it proceeds onward and may disappear—but at velocities above this limit the motion is unstable and a disturbance of the flow in the front part of the tube increases as it progresses onward until the whole mass appears in turbulent motion. When examined by proper means, such as a succession of electric sparks, this turbulence is shown to consist of rapid whirling and vortex motions by means of which the slow-moving water at the sides and the rapid-moving water in the axis become thoroughly mixed and thus continue down the tube till the motion throughout the whole section of the tube becomes slower and very nearly uniform at the limiting velocity upon which the system of rectilinear parallel motions then again begins and continues until some extraneous disturbance again throws it into a new series of whirls. In this process the consumption of energy in overcoming the viscosity is made to diminish the general velocity of the fluid until the latter finally reaches the limit of stable flow and continues on from that point down the tube steadily diminishing in velocity and without any more whirls introduced by instability, since instability does not exist at such diminished velocity.

7. These turbulent whirling motions are one form of the discontinuity of motion first studied analytically by Helmholtz and Kirchhoff. What the precise nature of the whirls will be must depend upon the nature of the obstacles or disturbances within the narrow tube, but the possibility of their origin and existence and continuance depends upon some relation between the co-efficient of viscosity of the fluid, its co-efficient of slip on the sides of the tube, its velocity and the diameter of the tube, the latter is, of course, the diameter of the largest vortex that is any way possible within the tube. Reynolds's experimental results show that the instability begins, or the limit of stable flow in the tube is reached, and a continued series of discontinuous whirls become pos-

sible within the tube, when the velocity of flow attains a certain critical velocity which in the case of water is expressed by the formula

$$V_c = \frac{\mu}{B\rho c}$$

where μ is the value of the co-efficient of viscosity of water at a given temperature; c is the diameter of the tube in metres; ρ is the density of the water at the given temperature; B is the co-efficient 278 as determined by Reynolds for cylindrical glass pipes of ordinary smoothness and symmetry and for Manchester hydrant water at the temperature 0° C. In general the critical velocity is proportional to the ratio

$$\frac{\text{Viscosity of the fluid}}{\text{Density of the fluid} \times \text{diameter of tube}}.$$

In order to apply Reynolds's results to the air we find the ratio viscosity divided by density which for water at 0° C. is $\frac{0.018}{1.000013} = 0.018$ but which becomes for air at 0° C. and 760^{mm} pressure $\frac{0.000188}{0.0012932} = 0.15$, or in other words in tubes of the same diameter the critical velocity for air is eight times that of water, or inversely, if the velocity of the flow of air and water is the same and the limit of stability be already attained in the tube of water then the limit will also be reached in the tube of air if the latter has eight times the diameter of the former tube; thus Reynolds's experiments give the following results for water at 0° C.:

Tube glass	Diameter meters.	Critical velocity (meters per second).
1	0.0268	0.7
2	0.01527	1.4
3	0.007886	2.05

Therefore, for tubes of the same size as these, the critical velocities for air will be respectively 5.6, 11.2, and 16.4 meters per second; or for air velocities of 0.7, 1.4, and 2.05 meters per second the critical diameters of the tubes would be 0.2144, 0.1222, and 0.0631 meters, from which we see that the critical velocity diminishes rapidly as the diameter of the tube increases. Although these observations by Reynolds relate to tubes whose walls are rigid yet they give us a first approximation to the conditions that attain in the free atmosphere where the ascending columns have flexible instead of rigid boundaries, and show us that the faster a narrow stream of air flows through surrounding quiet air the smaller must the diameter of the stream be if it would preserve its linear motion and integrity and not be thrown into whirls whereby it becomes mixed up with surrounding air.

8. A cylindrical stream flowing in straight lines through fluid of the same material is technically known in hydrodynamics as "a jet," and so long as it maintains a steady motion, offers analogous phenomena to that which occurs when flowing within the rigid boundaries of the glass tube, but just as within the tube certain disturbances may arise introducing vortex motions so also with the column of air (especially when surrounded by liquid or gas, and, therefore, having boundaries that are no longer rigid but extensible not to say elastic) it is common for disturbances from the outside to penetrate its bounding surface and cause the fluid within to assume a variety of motions.

The difference between motion within a tube and that in a region whose boundary is a surrounding layer of gas is illustrated by the ascent of warm air, smoke, etc., through a chimney. So long as the ascending air is within the chimney its motion is vertical and the roughness of the chimney retards horizontal gyrations and facilitates rotations in a vertical plane; but no sooner does it reach the free air, no matter how still the latter may be, than it enters with great freedom upon its rotation around a vertical axis; the latter very soon becomes inclined and even horizontal and the ascending air goes through a system of corkscrew contortions that end in its complete disintegration into revolving fragments causing its final disruption as an individual column and its dissipation through mixture with the surrounding air until it spreads into the umbelliform figures shown in Figs. 7 and 30, and which have been studied by Vettin, Oberbeck, and others.

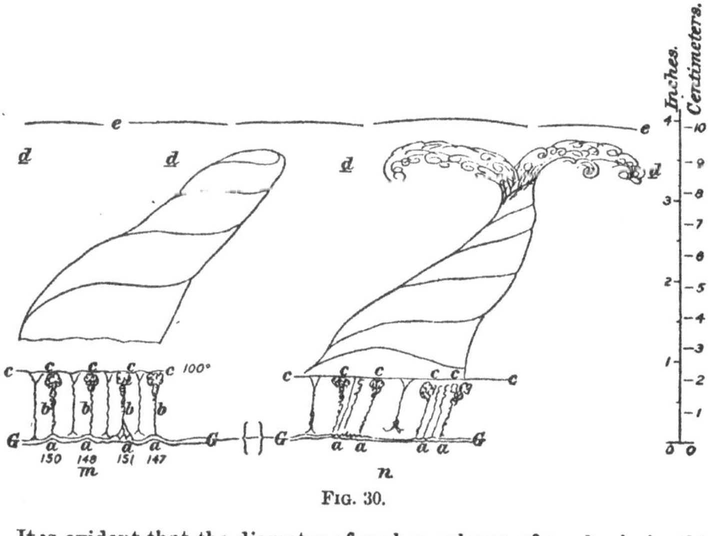

FIG. 30.

It is evident that the diameter of such a column of smoke is in this case too large in proportion to its velocity, hence the ease with which it is thrown into whirls and discontinuous motions, but it will eventually be spread into umbelliform at the top, where its velocity diminishes so that viscosity becomes the principal and finally the only resisting force.

9. As soon as the surface of the ground becomes overheated in the morning by solar radiation or otherwise, and ascending currents begin, we have occasion to apply the ideas drawn from the preceding illustrations.

At the immediate surface of the ground, or of the vegetation and waters that cover it, we find very great differences of temperature between closely adjacent surfaces, differences that have as yet never been directly observed but which all reason and analogy teaches us must frequently amount to many degrees. By reason of such temperature differences these innumerable points or minute surfaces, by heating the air in contact with them hotter than the surrounding air, give rise to corresponding little streams of ascending hot air whose upward relative velocities may be anywhere from 0.001 to 1 meter per second.

These minute currents mix with the alternating denser air in contact with them and give rise to the wavy motions that the observer sees when he looks at distant objects through such a mass of air. In general these currents start upwards as little streams of viscous air in steady, slow, rotary motion, but they soon acquire so rapid an ascent that their rotary motion is unstable and very soon the whirls initiated by outside disturbances break up the columns causing eventually a complete mixture with the adjacent cooler air; thus at the height of a few inches the ground is covered by a layer of more nearly uniform temperature, high indeed but much cooler than the average temperature of the surface of the ground and still cooler than the temperature at the hottest points at which the minute ascending streamlets began.

The accompanying diagram (Fig. 30) illustrates such a stream beginning at *a a*, hot points on the surface of the ground, ascending rapidly in direct steady motion to *b b* with a nearly uniform temperature, which in midsummer on the hottest soils may be 150° Fah.; then by its whirling motions rapidly broken up and by conduction lowered in temperature until, on reaching the points *c c* and breaking into smaller fragments or spreading in thinner layers, it has become thoroughly mixed with the minute streams descending from above and attains the uniform temperature which prevails at an inch or two above the ground along the level *c c* and which in summer may easily be 100° Fah. A similar process takes place over neighboring portions of the surface of the earth as at *m* and *n*; owing to the differences of rocks, earth's vegetation, or dampness the average temperatures of the surface and of the resulting uniform layer are slightly different at *m* and *n*.

In the process just described the buoyancy of the streamlets which is due to their temperature and consequent expansion is the ascensive force; that is to say, it is a difference of gravitation or weight of two neighboring portions of air that does the work of elevating the small masses of air and of overcoming the attendant resistances. Now a perfect fluid, free to assume a surface of least resistance, and at absolute zero of temperature, will pass through another perfect fluid mass of indefinite extent without suffering any resistance from the opposing external pressures or inertias of the moving masses; therefore the rapid diminution of temperature in the mass that is ascending from *a* up to the uniform layer, *cc*, is due simply to the mixture with the descending cooler air, and to the fact that the thermal energy of the ascending streamlet is consumed in the molecular work of overcoming viscosity; evidently the work done in the expansion of the volume of gas is inappreciable in the cases we are considering of layers within a few inches of the ground.

10. The spots of earth *m n* (Fig. 30) over which the air may be considered as having a uniform temperature will generally cover a very few square inches or feet; but the neighboring layers *c c* now act upon each other precisely as did those at the immediate surface, the warmer ones ascend and are very soon again completely mixed with descending air until new and broader layers *d d* are formed of uniform temperature at the height of a few yards above the earth's surface (see Figs. 30 and 31); the ascent of these larger streams is slower because the differences of temperature are smaller, and the changes of temperature within them as they rise are now due principally to radiation, to absorption and conduction, and to the intermixture of other air, and not to the overcoming of viscous resistance. A slight amount of heat is consumed in the mechanical expansion, but this can only be small, since we are now considering cases where the ascent is but a few feet at most.

11. The astronomer and geodesist are accustomed to find that at a certain hour in the morning and again in the afternoon the optical condition of a long horizontal stretch of atmosphere is exceedingly steady; thus at Poulkova a telescope, on the observatory hill at *p*, Fig. 32, directed horizontally to the gilded domes 12 to 20 miles distant, so that the line of vision is about 80 feet above the cultivated plain, shows in clear weather that up to 9 o'clock in the morning the air is full of streams of hot rising and cold descending currents so that objects appear ill-defined, blurred, faint and jumping both horizontally and vertically, but about this hour the troubled vision improves and for a few minutes the air is perfectly steady and vision perfectly satisfactory. This moment occurs earlier in the day and lasts for shorter intervals in the summer time, when the sun is more powerful in its heating effects; it also occurs earlier in the day at more southerly latitudes and earlier when the line of vision extends over dry plains easily heated than when it extends over water. This epoch evidently marks the moment when the mixture of hot and cold currents has extended up to the layer between *p* and *d*, and which moment is determined by the following

considerations: During the clear night the layer of air *o m* has cooled to the temperature of static equilibrium or stability; the temperature of the layers below *O D* being lower than that of those at *m n* the buoyancy of the lowest layer is increased by the solar heat, it rises and mixes and the process goes on until the higher temperature has attained to the layer *o;* it is only after the layer *n m o* has become thus gradually warmed up to a certain limit that it then begins, if it is still further warmed up, to invade the warmer air *o o′ D D′* above it.

The period of good vision in horizontal lines corresponds to the time during which the stratum *POD* is enough warmer than the air below it and colder than the air above it to be held in convective neutral equilibrium so that it can neither pass downward nor upward and must, therefore, be free from convection currents. Therefore, the duration of time from sunrise up to 9 a. m., or the moment of good vision, show how long time is consumed in warming up the lowest ground stratum of earth and the air above it and in evaporating the moisture at the surface until the air below *o m* has attained the needed rarity. We may compute how much heat is consumed in this operation. In the evening the operations are reversed, the cooling by radiation at the surface of the soil and the warming by insolation balance each other at 5 or 6 p. m., and there is a half hour or less during which the vision again becomes satisfactory because the horizontal layer below *o* has cooled so that it can not ascend and those above *o* are still so warm that they cannot descend.

The rate of diminution of temperature in dry air compatible with static equilibrium is 1.89° Fah. for 100 feet, or 3.42° C. per 100 meters of elevation or anything less than this, while any rate greater than this is proportionably favorable to convection. When convection has once set up the ascending currents cool at the rate of 0.98° C. per 100 meters ascent or 0.54° Fah. per 100 feet. If they cool faster than the existing rate of diminution in the static layers they will soon cease rising, and if their rate of cooling is slower than that of the static layer they will increase in buoyancy and rise faster until the process of cooling by mixture becomes able to check their ascent, when they finally spread out horizontally and cease to rise.

12. The process of formation of ascending warm spiral columns, which begins immediately after sunrise, goes on until, within an hour in clear weather or two hours in foggy weather, there are formed atmospheric regions of considerable extent, say 100 feet deep and a mile square over prairies and oceans, but of a much smaller size over hilly countries and cultivated lands, within which a temperature prevails that is nearly uniform for any horizontal stratum and slightly higher than that in corresponding neighboring regions. These large areas (*e e* Figs. 30 and 31) in turn acquire ascending movements, but the rate of ascension is generally slower in proportion to the horizontal dimensions of the areas; the preceding table shows that for a rising

column 300 metres in diameter the critical velocity for stable motion of air is very small. The ordinary atmospheric currents undoubtedly have greater ascensive velocity than the 0.00005 meters per second given by that table, and, although we can not ordinarily see them, yet we must recognize the existence in our atmosphere of slowly ascending revolving streams of air of large dimensions intermingled with others equally as large but descending, and both having as they move slow horizontal rotations that tend to cause them to break in pieces and mix together.

In this motion on a large scale the diminution of temperature with ascent is principally due to the mechanical cooling of the large mass of gas expanding against pressure; and to a less extent to mixture and conduction, and to a still less extent but still appreciable to radiation and absorption while the viscous and the inertia effect remain inappreciable.

This motion is the process that Espy first grasped as an important meteorological principle; that Sir William Thomson mathematically treated of as convective equilibrium, and that is now recognized as the most important application to meteorology of the laws of thermodynamics.

13. Although these large ascending currents are generally invisible, and especially so in such dry weather as prevails in the United States during northwesterly winds, yet the upper portion of the currents may become visible by the formation of clouds when they carry up enough moisture within them, so that Espy described cumulus clouds as being the visible summit of an ascending invisible column; that such ascending currents exist over all heated regions must be conceded on general grounds as inevitable owing to the differences of temperature at the earth's surface.

The general drift of any one such current becomes apparent to the eye when we consider the flight of buzzards and other carrion birds; such a bird may often be watched during a whole day, in the course of which he passes from one side of the horizon to the other over a horizontal distance of fifty miles or more, wheeling in steadily diminishing spirals alternately into and out of successive odoriferous regions as he traces down to its origin the long train of scented air that has been carried from decomposing flesh up to his height by slowly ascending currents of warm air.

These large masses of air rise slowly during cloudy weather, when their relative buoyancies are small, and also slowly during certain hours of the day; they generally cease in the lowest strata between midnight and sunrise, but may still continue in the strata just above until these upper strata lose their heat. Under other atmospheric conditions, however, they may rise quite rapidly, the extreme velocities being found inconnection with the formation of thunder-storms and tornadoes.

14. When the ascensions are slow as compared with their diameter

and horizontal movement the rotations and discontinuous motions by which they tend to be broken up are correspondingly weak and the large cumuli are formed.

When the ascensions are rapid and the horizontal movement also rapid the ascending currents are more liable to be broken up into smaller separate masses of rising air from which only scud and small cumulus clouds can then be formed.

This latter is a condition that frequently prevails when the air is cool and dry but the ground comparatively warm; thus at Washington during cool, dry northerly and westerly winds in the summer time we frequently find numerous small cumuli passing over us, each of which represents what had an hour before been a small mass of moist hot air close to the earth's surface but many miles to the northwestward. Similar small cumuli occur at Washington during the cold, dry northwest winds of winter, but the temperature and moisture are then lower and the clouds smaller.

In general the stream of lighter air ascending from the region *g g d d* (see Fig. 31) although renewed by the inflow and the warming up

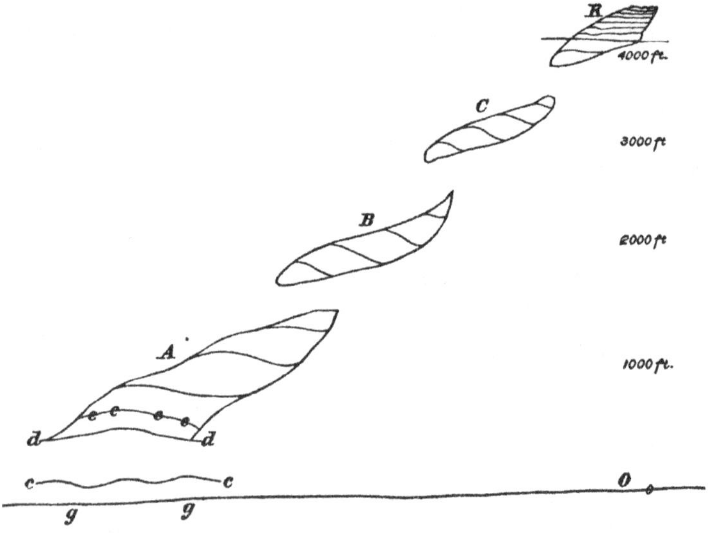

FIG. 31.

of other air, rises too rapidly to maintain the continuity of so large a mass, and in its rapid rotation is broken into a series of masses *A B C*, each of which becomes a smaller circular or cylindrical vortex and rises until, as it cools to the dew point, it forms a separate cloud as at *K*; thus an observer to the leeward of *A* (Fig. 31) observes a series of cumuli floating over his zenith, whose distance apart represents the time required by any mass to rise to the cloud level while its preceding cloud mass moves forward.

Let D = horizontal distance of centers of small cumuli in feet per second.

W = horizontal velocity of clouds in feet per second.

V = vertical velocity of ascent.

$T = \frac{D}{W}$ = time elapsing between the arrival of successive mass centers at cloud level.

$VT = D\frac{V}{W}$ = vertical distance of cloud masses above initial level.

In considering the clouds over one's head the observer has to remember the horizontal and vertical distances that correspond to a given angular slope or gradient; thus a horizontal movement of 10 miles hourly or 4.5 meters per second, and a vertical movement of 1 mile per hour or 0.45 meters per second, gives a gradient of 1 in 10 or 5.4°. A horizontal movement of 45 miles per hour and a vertical movement of 1 mile per hour gives a gradient of 1 in 45 or 1°.2. The latter corresponds nearly to the buoyancy gradient in ordinary storms, and the former to the buoyancy gradients in the early stages of formation of a thunder-storm; for example, on May 17, 1889, a delicate cirro-stratus cloud is observed at Washington, at 1 p. m. moving east, its striæ trending to the southeast and northwest, and at 1.50 p. m. trending east, and 2.10 p. m. trending east-northeast. Otherwise the sky is clear all day. The air forming this cloud may have started from the ground at some point in the mountains to the west of us at 9 a. m. or earlier (possibly even on the previous day), and has been slowly rising and thrown into these wave-like rolls by the action of adjoining currents combined with its own slow rotary motion. The morning weather map shows a general southwest wind below and southerly wind above, but the higher cirrus cloud that we are observing is moving as a body from the west to the east.

15. The formation of rising streams and clouds may start at any layer where its temperature and altitude are properly related, and this is determined by studying the temperature in a section of the atmosphere east and west over any station, such as is illustrated by the diagram Fig. 33, which results from the annual means of the frequent observations made at Allahabad, India, for two years, 1886–'88, and is reproduced as originally drawn by S. A. Hill (see Indian Memoirs, 1889, Vol. IV, § 9); this was intended to show the temperature above Allahabad at any hour of the day and at any altitude up to 200 feet, at which the observations ceased; but the horizontal line or scale of hours of local time of the original diagram may be considered also to be a scale of degrees of longitude; thus, if it is 5 a. m. at Allahabad then simultaneously we should have 6 a. m. at the next hour line to the right, or fifteen degrees of longitude distance to the east, and 7 a. m. at the next point thirty degrees east of Allahabad. Therefore, the temperatures at the surface

of the earth and those at any height above the surface for any place to the eastward may be considered as shown by the figures on the curved lines of the diagrams; thus, all points at the longitudes and altitudes connected by the curved line of temperature passing through the hour point 7 p. m. have the same temperature 78° Fah. that prevails at Allahabad.

The slowness with which heat penetrates upward by convection and its greater accumulation in the afternoon hours in the atmosphere is easily seen from a study of the curved lines between 10 a. m. and 6 p. m.; on the other hand, the decided cold at low altitudes at night time and the warmth of the layer above is seen by studying the curves between 10 p. m. and 5 a. m. It is the warm air of this upper region, that has not been much cooled by nocturnal radiation, and which, by reason of the heating of the lowest strata by the early morning sun of the next day, is brought down by convection currents in the morning hours, that constitutes the dry, cooling, refreshing morning or day breeze. Ordinarily the air thus brought down accumulates a share of its heat day by day and its lower temperatures slowly increase up to a maximum when convection again takes place on a broader scale and to greater heights whence in time cooler, drier, or denser air is brought down and the warming process begins anew with it.

16. As illustrating the ascent of buoyant gas through a layer of greater density, and the inevitable occurrence of rapid rotary motions breaking up the stream of gas into discontinuous portions, may be cited the following case lately noted by myself: *a a* Fig. 34 represents the glass sides of a Piche Evaporimeter; *b b* is the small paper diaphragm separating the column of water within the tube from the outside air. An accidental microscopic aperture allowed a fine stream of air to enter evidently pushed in by atmospheric pressure since the space above *d d* contained only aqueous vapor. So long as *C* was sufficiently small the inflowing air formed a continuous tube from *C* to *e* and apparently flowed upward in straight stream lines; when the aperture became slightly larger a distinct spiral motion became evident in this air tube; and when it was a little larger still the spiral column was broken up at *g* into a beautiful series of bubbles of which the upper ones, owing to buoyant acceleration, rose faster than the lower ones.

17. The process of convection thus described is that by which the distribution of heat and moisture in the atmosphere is largely controlled during the day time. At night time the radiation between the upper and the lower strata on the one hand, and between the upper atmosphere and outer space on the other (which radiation exists during the day time but is masked by the effects of convection) becomes the controlling feature.

As the diurnal convection currents appear not to be very active in a vertical direction at elevations above 30,000 feet, although the horizontal currents continue to be of great importance, therefore, somewhere at or above this level there is a region where throughout the day (and another region where throughout the year) radiation, direct absorption, and even conduction are more important than convection; therefore, at and above these altitudes, the diurnal periodicity of the temperature of the air is very slight, depending principally upon the direct absorption by the rarified air (and by its occasional dust haze) of the solar and terrestrial radiations. The solar radiation sends to us both short and long waves, namely, rapid and slow molecular vibrations, of these the rapid ones are absorbed most greedily by the upper atmosphere which transmits them to us again, in a modified form, by a process of fluorescent-degradation, as blue light; the terrestrial radiations are rich in long waves or slow vibrations that are but little absorbed by ordinary clear atmosphere, and are almost completely transmitted through the thin air of the upper regions. Therefore, the diurnal period in the temperature of the upper strata depends mostly on its own direct absorption of the solar radiation, while terrestrial radiation passes through it unhindered day and night.

Finally as we ascend still higher the direct absorption of the short waves of direct solar radiation during the day-time becomes less important than the radiation from the atmosphere itself, and at this altitude temperature becomes constant throughout the day, and almost so throughout the year. I estimate this altitude at not much above 20 miles, but it will vary with latitudes and seasons and with the slightest change in the chemical and mechanical constituents of the atmosphere. Thus meteoric dust, volcanic vapor dust, or light gaseous compounds, ejected from volcanoes and factories, will affect the absorbing power, and, therefore, the temperature of this layer, and may produce barely appreciable temporary effects on the temperature and convection in the lower strata; indeed I see no reason why a secular accumulation of lighter hydrogen compounds should not be in steady progress at this level, such as may produce a slight secular increase in the absorption of solar rays, and in the temperature of the earth's surface, and, therefore, of the lower atmosphere. Every increase in the absorbing power of the air necessitates that the temperature of the earth and the absorbent must be thereby raised until the excess of its temperature above that of its surroundings will enable it to radiate its own heat at the same rate that it is receiving; therefore, it matters not whether the atmospheric absorption for long waves or for short waves be increased, the result in either case is the same, that each layer of the whole atmosphere must attain a higher temperature in order to maintain a steady temperature. On the other hand, if the upper layer be subject to a periodic or an irregular interjection of foreign material and corresponding variation in its absorbing power, there must be a corresponding variation in its own temperature and a less noticeable parallel temperature variation in the lower atmospheric strata.

18. The study of the distribution of temperature in the atmosphere is, therefore, divided into sections as follows:

(*a*) The lowest stratum, namely, from the upper surface of the ground or fog, up to the lowest cloud surface: throughout this region temperature is in the day-time principally regulated by purely convective action due to interchange of air in the dry stage, but in the night time this stratum is to be divided into two portions: the lowest near the earth, where convection is entirely wanting and radiation controls the temperature; the upper one constituting by far the greater portion of the stratum in which convection still goes on but is less active than in the day time.

(*b*) The stratum in which convection is effected principally by the latent heat of condensation or by the convection of air into and out of the rain stage; this stratum extends from the lowest cloud surface to the highest cirrus.

(*c*) The higher stratum whose temperature depends upon the equilibrium between the absorption and the radiation that goes on between this stratum, on the one hand, and the clouds, fog, earth and the air beneath, and the sun, moon, and stars above.

(*d*) The stratum wherein absorption, both of the solar and terrestrial radiation is less than its own radiation, whose temperature is essentially due to conduction and is sensibly that of such matter in the adjoining space as does not partake of the motion of the earth, either diurnal or annual. It may appear rather startling to assume that this last higher stratum is in contact with quiet gaseous matter outside; it has been customary to think that the earth and its atmosphere moves through a void space, assuming that otherwise the resistance to the earth's motion of even the lightest ether or gas should become sensible in the astronomical observations on the length of the year. This assumption seems to me scarcely necessary in view of the fact that hydrodynamics demonstrate that up to a certain velocity the movement of a fair-shaped body through a thin gas is resisted only by the viscosity of the gas.

Now the earth's atmosphere gives us a yielding fair-shaped surface on which the ether of space acts only by viscosity, but the viscosity of the ether is evidently zero, or almost zero, therefore, this form of resistance is inappreciable. On the other hand if there were imperfectly elastic gas in space it would not flow in behind the earth with sufficient rapidity, and it would thus leave a pressure in front of the earth unbalanced by an equal static pressure in the rear; but the velocity at which this action comes in play depends principally upon the elasticity of the ether and the fair shape of the earth, and both these are almost perfect, therefore the velocity at which the earth would begin to be affected by this form of fluid resistance is one far above that which the earth actually has in its annual orbit. It is only in the case of a few comets at their perihelion that velocities are exhibited great enough to develop a barely appreciable amount of this form of fluid resistance.

The above-mentioned fourth or highest stratum may at times let some of its coldest air descend, but this convection is an important feature only in great general changes that require months or years to complete their cycle; thus if the sun-spot phenomena causes a periodic change in the solar radiation and a corresponding change in the distribution of temperature throughout the earth's atmosphere it can easily happen that air which would otherwise have stayed at a high elevation shall periodically come down to the earth's surface.

The third stratum above mentioned is that from which some have supposed the cold air to descend that constitutes the cold waves that, starting east of the Rocky Mountains, flow southeast over the United States; similar cold waves spread from Siberia and Russia southwest over Europe, and from Thibet southeast over China and Japan, as also from the Andes eastward over the plains of South America: oftentimes, in fact, a portion of the dry cold air on the eastern slope of the Rocky Mountains is drawn toward the Pacific coast, and flows into the valley of the Sacramento and more rarely into the San Joaquin. It is possible that the abnormal low temperatures in our third stratum (*c*) would cause descending currents of air, but as such currents are heated enormously by condensation under pressure, therefore, practically it seems to be proven that such great depressions of temperature as would be required to produce our cold waves do not generally occur in this third stratum, but that, on the other hand, the cold air that constitutes these waves is simply the dry air in portions of the lowest or first stratum cooled by radiation toward the cold ground below it, and toward the clear sky above it, until it accumulates in deep layers in the long winter nights of the arctic regions and flows toward any region where ascending buoyant warm air offers it enough opportunity. Its flow is due to a slight gradient of pressure in the direction of the flow: it is not a cyclonic flow around a low area in the first portion of its course, although subsequently it generally becomes so. A cyclonic circulation owes its power to the indraught of a central region where clouds and moisture and temperature have conspired to produce a buoyant atmosphere; the flow of the cold wave begins while the cyclone is still far off and in its first stages of development; the lower layer of cold air will flow whenever and wherever the distribution of pressure is such as to facilitate the tendency due to its density, by reason of which dense cold air flows toward and under warm, moist, and light air.

During clear weather the clear air of the cold wave in its original region of high barometer experiences a continued cooling by radiation that is only partially counteracted by solar heating during the day time, but as soon as the cold wave has been covered by a layer of haze or cloud (formed in the moist air that the cold air has lifted up) then its own temperature rises rapidly. Observations show that there is always a layer of warm air sometimes rather high above the advancing cold waves, and it is likely that this is also true even in the northerly regions whence the cold air flows down upon us.

19. The general horizontal distribution of heat in the lower atmosphere is ordinarily shown by isotherms, but the vertical distribution is not so easily observed or presented for study, and yet is specially needed in the study of simultaneous observations, since it is the changes of simultaneous conditions at successive moments that produce the disturbances in steady motion that constitute the difficult portion of the problems of meteorology.

Approximate curves of simultaneous isothermal lines or "synthermals" in the lowest air stratum were first given in 1867, by Hennessey (Royal Irish Academy, Transactions, Vol. XXIV—Science). Happening to see this memoir immediately after its publication, I saw that Hennessey had taken the step desired by Espy when he urged simultaneous observations for storm studies, and that his synthermal curves gave us the means of introducing into Ferrel's formulæ the *sine* and *cosine* terms needed to develop the diurnal period of storm motions and of the variations of barometric pressure; they afforded the sufficient basis for my decision to call for and publish only simultaneous observations in the Weather Bulletin of the Cincinnati Observatory For Hennessey's curves there can now be substituted better ones based on actual simultaneous observations made for the Signal Office throughout the northern hemisphere at Greenwich noon.

But we must superpose upon these curves, that hold good only for one stratum near the earth's surface, similar curves representing several higher strata if we would thoroughly appreciate the buoyant forces that are at work over the earth's surface. An extensive series of frequent observations of temperatures at various altitudes, or in lieu thereof some theoretical knowledge of the law of diminution of temperature with altitude is needed in order to attain this more thorough insight into the motions of the atmosphere. From an observational point of view the data recently published by the Meteorological Office at Calcutta, and above referred to in paragraph 15, are valuable as applying very closely to the hot and dry interior climate of the United States. If on the chart Fig. 33, copied from Mr. Hill's memoir, which shows the annual average distribution of temperature by hours throughout the lowest 200 feet of the air we follow up any vertical hourly column we shall find the temperature at any altitude to the nearest whole degree by noting the intersection with this vertical synthermal surface, and the fractions of degrees are given by interpolation between these curves, so that by following the horizontal line at any altitude we find the temperature at any hour of the day. The synthermal lines as here drawn are merely geometrical conveniences for interpolation and their physical reality in nature is not apparent until we introduce the following conception: assume that east and west of Allahabad (lat. 25° 26′ N., long. 81° 55′ E.) the earth and sky have all round the earth the same features as at that place so far as affects the temperature of the air, then similar observations on that parallel of latitude would have given similar local temperature curves throughout the circumference of the earth. Consequently the horizontal time scale of hours at Allahabad can be transformed into a horizontal and moveable scale of longitude of 15° to the hour, east and west of Allahabad. On this scale 7 a. m. standard time is noon at Greenwich or 5.28 a. m. at Allahabad.

On the scale of longitudes at the bottom of the diagram, Fig. 33, let the point marked *A* correspond to Allahabad at 81° 55′ and let the scale of longitude be considered as movable. Bring *A* to any given hour on the time scale of the plate and the zero will come opposite the corresponding time at Greenwich. The vertical line above *A* will show the distribution of temperature then prevailing above Allahabad; that above *O* will show the simultaneous distribution then prevailing at Greenwich, and so on for other longitudes. If the horizontal straight lines are imagined as bent around the earth so as to form the small circle of Allahabad, while the perpendicular lines become vertical to the earth's surface, or radial towards its center, then the curves will present the distribution of temperature in the atmosphere, at least for the hot, dry regions like Allahabad. A similar set of curves for the superposed strata, as also those for oceanic and for the continental regions, would enable us to present the normal distribution at Greenwich, noon, around the world with much accuracy.

The manuscript signal-service charts of international simultaneous observations, when combined into normal synthermal charts for months and years, will give invaluable indications as to the distribution of heat and the periodic disturbances in what would otherwise be a steady condition of circulation in the earth's atmosphere. The theoretical views expounded by Professor Ferrel enable us to see that the synthermal lines when carried around the world at the latitude of Allahabad, but at different altitudes above the earth's surface, will give a section somewhat as in the diagram in Fig. 35, which illustrates the slight variation at great heights and the higher average temperatures at any altitude over the eastern continent as compared with the western; the temperatures at sea-level adopted in this figure are those given on Hennessey's chart for the whole year, for latitudes 30° north, for Greenwich, noon.

We see here at once the beautiful hydrodynamic problem presented by the atmosphere, namely, the motions in a gaseous layer surrounding the globe, whose upper surface is of uniform temperature and structure except when slight changes occur which we neglect in dealing with the general problem, and endowed with uniform rotation but subject to a periodic wave of heat and expansion, and of evaporation and condensation; a slight viscosity in the lowest layers diminishing to almost nothing in the upper layers on account of their low temperature, resists the periodic tendency to internal motions; the interference day by day

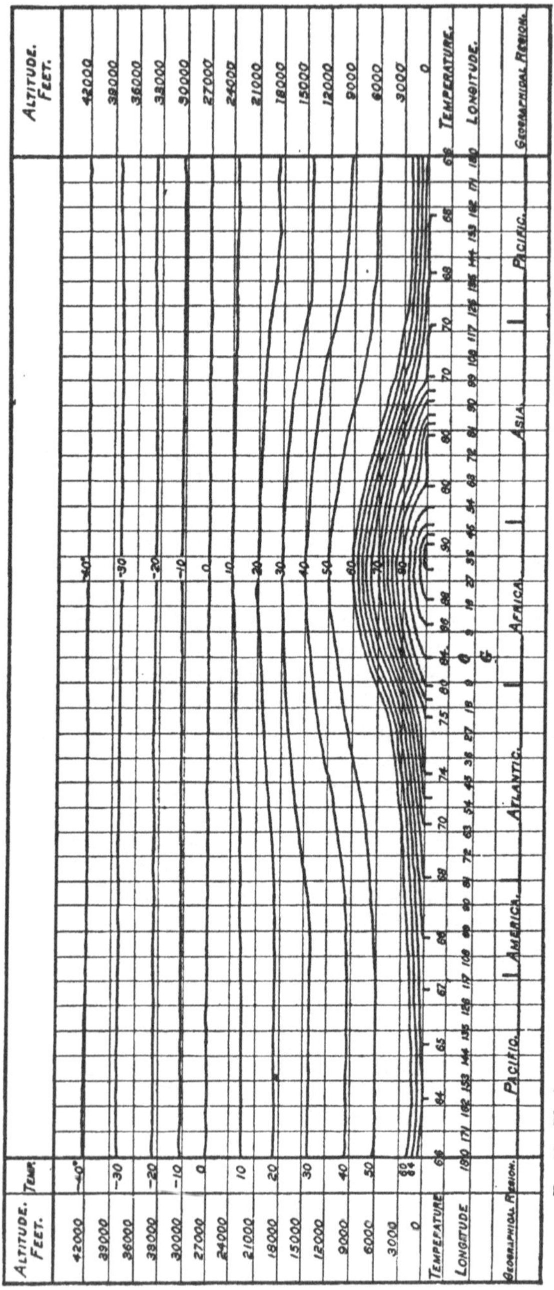

FIG. 35.—Ideal section of atmosphere around the earth at latitude 30° N., based on Hennessey's annual mean Synthermal Lines for Greenwich, 2 p. m.

of the new motions with what was left over from day before may also be neglected so that the periods of the perturbations consist of whole numbers of days or years; the motions of the lower strata consist of vertical ones directly due to heat convection and of horizontal ones principally due to the pressure due to the weight of the strata above. The motions of the lower strata come under the influence of the resistances of mountain chains and continents; of the former the most interesting is that of the Rocky Mountains and Andes; the flow of the lower strata over this mountain chain, which reaches up through at least half of the depth of the cloud stratum, introduces an immense disturbance deflecting the northeast trades of the northern hemisphere to the south and the southwest winds toward the north, which deflections are felt to a great height in the atmosphere and result in general gyratory movements of large areas about their centers, which themselves have a general movement about the north pole. There are thus produced those irregular fluctuations in storm tracks and cold waves that eventually bring about the irregularities in the distribution of rain-fall that constitute floods and droughts. The general circulation and the interconnections of temperature, pressure, and wind were first given by Ferrel in 1858, and lately more elegantly but with very similar results by Oberbeck.

As the annual radiation from the sun does not cause a sensible increase in the earth's temperature from year to year one must conclude that the loss by terrestrial radiation is fully equal to the amount of heat received. Now, a surface normal to the solar rays receives 3.0 calories per minute per square centimeter, and the quantity of heat received between latitudes 30° north and 30° south vastly exceeds that received on the two polar sides of these parallels, consequently the heated equatorial air must flow north and south into regions where it can cool sufficiently to balance this excessive reception of heat at the equator. The general co-efficient of emission of the atmosphere into space, which co-efficient is effective almost uniformly over the whole surface of the earth, must be sufficient to dispose of all the heat that the earth receives from the sun or on the average about 0.404 calories per square centimeter per minute (adopting Langley's 3.0 as the solar constant). Thus we see that the province of the high strata flowing from the equator toward the poles is entirely similar to the province of the Gulf Stream in maintaining the steady temperature of the ocean, and the fact that this atmospheric flow takes place so high up as to be almost inaccessible to our means of observation, must, on the one hand, render it impossible to at present predict general changes of seasons, but, on the other hand, must stimulate us to discover some method of observing what is going on so high above us. The influence of these great upper strata was felt from the beginning of my signal-service weather predictions when it was my daily custom to note the changes that occasionally took place in the general trend and velocity of motion of high and low centers. Very frequently after a number of storms had pursued parallel paths and with similar velocities it was observed that a new set of storm centers began pursuing a different system of paths, affecting the distribution of rain or heat and cold in a manner entirely different from those of the first group. Such changes were associated with droughts, floods, cold waves, and other atmospheric seasonal and climatic phenomena, and were evidently dependent upon conditions existing over the northern globe on a scale too large to be studied on the weather map of the United States. Therefore, early steps were then taken toward securing a daily map of the whole globe, and the study of the map of the northern hemisphere has sufficed to show that these changes in types of storm paths or average weather conditions are largely due to the density, *i. e.*, dryness, and temperature, and the motion of the upper currents of the atmosphere. A slight defect in the pressure or density of that layer at any point, or the combing of the general current as it swiftly flows over the summits of mountain ranges, or a wave-like action, such as represented by the sine and cosine terms in the harmonic series that results from the development of the dynamic equations, will either of them explain the existence of regions of fast and slow storms, or of regions of storm tracks differing in latitude, or of regions where storm centers have a periodic tendency to originate.

20. Diurnal distribution of vapor by convection in the lower air. The distribution of vapor as such, both horizontally and vertically, is wholly by the convection currents that also distribute the heat and it, therefore, depends like these, ultimately, upon the overheating of the lower air. When convection diminishes then the quantity of vapor in the upper stratum may become stationary or even diminish if the upper air contains enough to allow precipitation. Now the vapor in a given volume of the lowest stratum is diminished by convection by reason of the admixture of the drier air brought either from colder places on the earth or down from the stratum above; the down flow of air is not likely to diminish temperature, but is almost certain to diminish the humidity: We have in fact a lowest stratum at the earth receiving wxy units per hour of moisture, where w is a factor depending upon the wind, x is a similar factor depending upon the temperature, and y depending upon the moisture possible to be evaporated from the surface of the soil; we have also an upper stratum containing a grains less of moisture per unit volume. Therefore, the moisture contents of the unit weight of intermediate air is $ma + nwxy$ where m and n are the relative proportions of the upper and lower air in mixture.

Evidently, for the very lowest stratum, convection is very active from 9 a. m. to 1 p. m., but the down-coming air is already much nearer its dew-point than that which is brought down in the afternoon from 1 to 5 p. m. By 5 p. m. convection, in so far as it exists, has ceased to carry up air from near the surface of the ground and similarly has ceased to bring other air down to that level—it both begins and ends at a level considerably above that of the earth's surface; the occasional puffs of air that are felt show when a mass of denser air happens to be borne along over us. In the morning small convection currents extend from the surface of the ground rapidly upwards; each such current ceases when it has developed the umbelliform structure, before described, and its buoyancy has been consumed by radiation and by mixture with cooler air. In the course of the afternoon when convection extends further up into the dry air and evaporation at the surface of the earth has diminished by reason of its diminishing temperature, so that the larger proportion of the air at that surface is that which has not sufficient buoyancy to rise again, then we should expect to find the vapor tension less in the lowest strata than in those just above, or at least the rate of diminution of vapor like that of temperature as we go upwards becomes a minimum at some definite altitude.

The following table shows the results of the observations made by Hill at Allahabad for one and two years and are all that I know of that will serve to illustrate the diminution of vapor tension with altitude at different hours of the day:

Vapor tensions—Annual means for Allahabad, India.

Hour, local time.		Two year means.			One year means.	
		Ground shelter.	6-foot shelter.	46-foot shelter.	104-foot shelter.	166-foot shelter.
Midnight	0	0.604	0.598	0.564	0.532	0.513
	1	.587	.596	.569	.540	.521
	2	.586	.595	.570	.512	.524
	3	.584	.593	.571	.544	.525
	4	.581	.590	.570	.542	.524
	5	.579	.587	.570	.543	.524
Sunrise	6	.583*	.591*	.572*	.548*	.527*
	7	.591	.600	.573	.557	.537
	8	.596	.609	.572	.563	.542
	9	.594	.612	.564	.561	.541
	10	.587*	.605	.550	.550	.530
	11	.573	.589	.533	.533	.516
Noon	12	.556	.572	.516	.513	.497
	13	.540	.555	.499	.497	.482
	14	.528	.544*	.489*	.485*	.469*
	15	.521	.537	.487	.479	.463
	16	.526*	.542	.494	.478	.460
	17	.549	.558	.501	.483	.461
Sunset	18	.574	.582	.528	.492	.468
	19	.593	.599	.539	.499	.476
	20	.598	.608	.546	.506	.482
	21	.597	.607	.551	.512	.489
	22	.593*	.605*	.555*	.520*	.496*
	23	.590	.601	.560	.526	.505
Midnight	0	.604	.598	.564	.532	.513

* These are the standard observations made by eye to check the records of the thermographs.

21. We shall have occasion to make use of the figures representing the relative density of the air, and that, too, not only near the earth, but at great altitudes. To this end the accompanying tables are prepared as an extension of those of Landolt and Börnstein. The general formula for the density of the air and the values of the constants that we shall adopt are as follows:

$\rho_0 =$ 0.001293052 grams per cubic centimeter, the density of dry air containing an average amount of oxygen, nitrogen, and carbonic acid gas, at the standard temperature of freezing water and the standard pressure P of one atmosphere (760^{mm} mercury at 0 and at standard gravity).

$\rho =$ the actual density for any temperature, moisture, pressure, and gravity.

$g =$ the actual and g_0 the normal force of gravity.

$b =$ the observed pressure or the barometric height reduced to standard gravity and temperature.

$e =$ the observed tension of vapor expressed as a barometric height under standard gravity.

$h = b - \frac{3}{8} e$.

$a = 0.003670 =$ the co-efficient of expansion of air for temperature.

$$\delta_{tbe} = \frac{0.001293052}{1 + 0.003670\, t} \times \frac{b - \frac{3}{8} e}{760} = [\, t\,] \times [\, h\,]$$

When b, t, and e have been observed, as at all meteorological stations, there remains only to compute the density by the use of these observed values, and this is easily done by the use of tables based on the above formula, but when these elements have not been observed as for air at high elevations, it is necessary then to in some way ascertain, by the most plausible hypotheses, the approximate values of the pressure, temperature, and moisture prevailing at this elevation. This brings us face to face with the hypsometric formula and the uncertainties that attend its use.

CHAPTER V.

CLOUDS AND THEIR HORIZONTAL MOTIONS.

The relative importance of the slow ascent of a general current up the general slope of the continents, and which I will call a gradient due to topography, and the steeper or even vertical ascent due to buoyancy or density of the cloud is shown by its shape, proportions, and inclinations. Some careful measurements of the altitudes of clouds by the Swedish observers have shown that some clouds are ascending while others are descending. The vanishing-point apparatus described in my treatise on instruments and methods (§ 138) shows at a glance, and in a very general way, whether the clouds of a given region of the sky are ascending or descending, or in general moving parallel to each other. But in the absence of exact measures, and for the general use of observers, a careful study of the appearance of the clouds will frequently show the character of their motions.

Clouds are usually studied with a view of answering the query: "Will it rain soon?" But our present study is to enable us to learn from them somewhat as to the relative motions of the cloud or rising mass as a whole, and of the air within as compared with the air at the boundary of the cloud, or the orographic and the buoyant gradients respectively. These are, I believe, distinctively shown by the general shape of the boundary of the cloud and may be classified according to the relative gradients, which are approximately as shown in the following schedule:

Cloud figure number.	Orographic gradient.	Buoyant gradient.
36	Rapid	Steep.
37	Steep	Gentle.
38	Gentle	Do.
39	Inappreciable.	Rapid.
40	......do	Gentle.
41	......do	Inappreciable.

If we pass from the lower cumulus and stratus of the preceding schedule to the upper cirrus and cirro-cumulus, and even to the cirro-stratus haze, the study of their forms will also give an indication of the vertical and horizontal movements going on. Thus in the cirrus, frequently described as "Mare's Tails," "Pele's Hair," "Polar Bands," and other fibrous or filiform structures, we see evidence of the formation and disruption of the ascending cylindrical vortices in the thin air of the upper regions. The cirrus clouds themselves (see Figs. 42 and 43) represent the long streams and the curved surfaces of separation of otherwise invisible rolls and vortices where condensing vapor becomes temporarily visible, as illustrated in such experiments on vortex motions and stream lines as can be made with smoke on a small scale in the laboratory.

The cirro-cumuli that cover the sky with regular patterns, known as mackerel sky (Fig. 44), Noah's Ark (Fig. 45), etc., show small clouds remind one of that the standing waves and ripples produced by the flow of two strata of water over a rocky bed, or in the case of the atmosphere by two strata of air, in parallel planes but in different directions, thereby producing standing waves at the boundary surface between them, as in Figs. 46, 47, and 48.

Now these standing waves in water and in the air are in general due to the system of rolling or vortical movements, of which a very fine example can be seen on looking down from the deck of a steamer close alongside the hull; by attentively studying the water within a foot of the surface of the hull we shall find that while the water at a distance is quiet so far as the boat is concerned, yet close to the boat there is, as it were, a cushion of water in a state of violent commotion; the particles adjacent to the hull are pulled rapidly along with it in its motion, the layer next to them follows more slowly, those a foot away are not only not pulled along, but are as it were pushed back by the pressure of the outside water in order to take the place of those which have been pulled along, consequently a series of little vortices is formed, as in Fig. 49, and we see that the so-called cushion of water is a mass

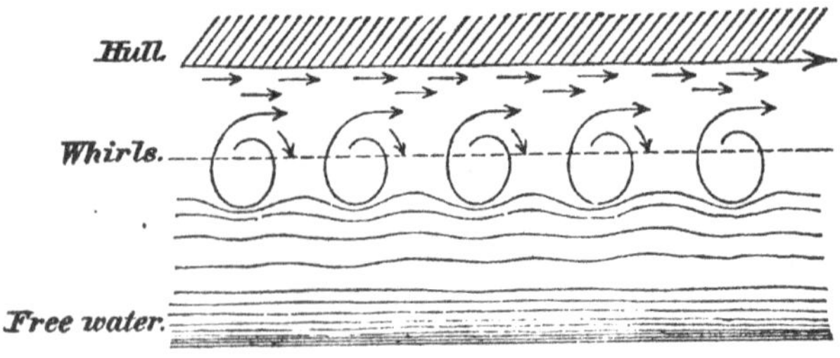

FIG. 49.

of rolling rollers whose centers, and, therefore, whose masses, move along with about half the velocity of the boat, and are continually being left behind it. This phenomenon is not seen for slow-moving boats, but only for those whose velocity exceeds a certain limit; this limit is similar to the one determined by Osborne Reynolds for the flow of water in pipes, namely the limit, which depends upon viscosity, at which steady motion becomes difficult and discontinuous motion becomes possible and easier. The thickness of the cushion of water depends upon the velocity of the boat, the viscosity of the water, the roughness of the hull, and the shape of its water lines; as the whole mass of water involved in this layer of vortex motions is moving along with the boat at one-half its velocity, therefore, in moving twice its own length, the boat has communicated to this mass of water an amount of energy represented by the motion of this mass at one-half the velocity of the boat, and this immense consumption of energy constitutes a very large portion of the power consumed in attaining any given speed.

To pass from the boat to the cloud motions, we see that the adjacent layers of air whose motion is in parallel planes but different directions, as shown in Fig. 50, must have between them a similar set of atmospheric rolls in each of which the separate particles of air describe ellipses or circles in planes that are nearly vertical and symmetrically arranged, forming long rolls, as in Fig. 42. The upper layer of air

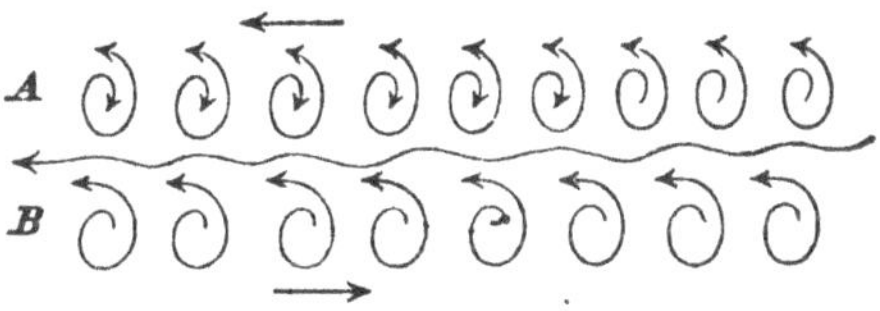

FIG. 50.

A may be dry and cold, while the lower *B* is warm and damp, or *vice versa*; the former unstable, the latter stable; but either arrangement is equally likely and, so far as balloon observations tell us, equally frequent, since the aeronauts frequently describe the passage from one side to the other of a horizontal layer of abnormally warm or cold air, therefore, the effect of the circulation within each of these invisible rolls is of two kinds: first, the warm moist air of one layer may be brought into contact with the colder, drier air of the other and form parallel streaks parallel to the rolls of very delicate haze; or, second, the moist air of the lower stratum when carried up to the top of its circular course may, by expansion, be cooled just enough to form a delicate cumulus cloud; of course this cloud will then represent merely the locus of the air in that portion of its circulation in which its temperature is below its dew-point, for when carried beyond this point and carried down on the other side, its moisture evaporates and becomes invisible, a process that will be illustrated in the next paragraph by illustrations of clouds formed in this way on a much larger scale.

The movements of the individual cirro-cumuli are thus seen to be the same for all those in a given set and are the resultant between the movements of the upper and lower currents. Usually the individual small cumuli are seen to be dissipating on the one side while forming on the other, and their apparent movement along the sky is not exactly that of the air, but (like wave motion on the surface of the water) indicates merely the propogation of the form of motion, thus, in Fig. 51,

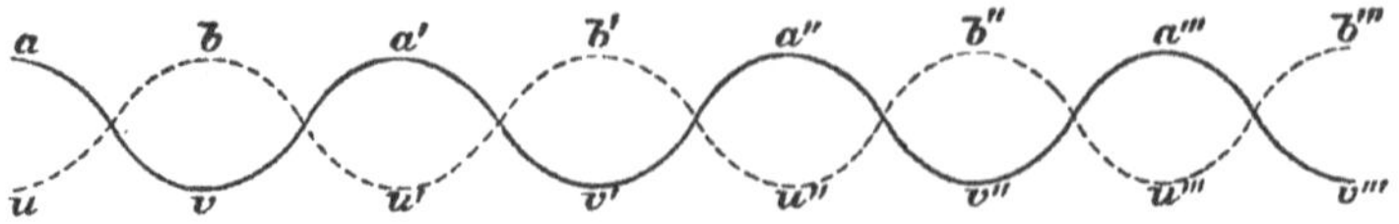

Fig. 51.

let *a a' a''* be the tops of the waves represented by successive cirro-cumulus clouds, after a few minutes the matter at *a* has been brought down by its undulatory movement to *u u' u''*, and has become invisible, while the air that was at *v v' v''* has been elevated to *b b' b''* and has become visible as small clouds, and we say that the cirro-cumulus has moved from *a* to *b*, a motion which we now see is not the horizontal movement of the mass of air, but the horizontal progress of a system of waves; the relation between the progress of a group of waves as a group, and the progress of the individual waves has been developed by Lord Rayleigh, and applying his conclusions for water to the atmosphere we see that the movement *a b* of the individual cirri takes place with twice the velocity of movement of the whole group constituting a definite form, such as a Noah's Ark or a patch of mackerel sky.

Dissolving cirrus, that is to say, a general disappearance of the cirro-cumulus or other forms may be produced either by the heat of the sun, or by the mixture with drier air, or by the general descent of the mass on whose border the cirrus is formed, and these three methods are precisely parallel to the similar changes that take place in the lower cloud region. The most frequent and important changes are those that result from the increase or diminution of the supply of ascending or moist air.

As a rule the striated and cumulus forms of cirri (see Weilbach on "Forms of Clouds in Northern Europe," Annales Bur. Cent. de Met. de France, 1880, Part I) are in our latitudes most numerous in the morning and early afternoon, but disappear during the late afternoon and evening hours. (See Espy, Fourth Met. Rep., 1859, pp. 49 and 171.) Their formation is due to the rising currents in the higher part of the atmosphere, and ceases with a declining sun precisely as is the case with cumuli in the lower strata. Espy was of the opinion (Fourth Report, p. 171) that the cirri are all made out of tops of rain-clouds, being the remainders of the clouds after the clouds have ceased raining, and Clement Ley has advanced the same view. But having frequently seen the cirrus form without any connection with rain-clouds I should say that they may also frequently represent the ascending currents and mixtures that take place in the thin upper air. In other words, ascending currents may start from a high plane in the atmosphere just as they start from the earth's surface, and such currents may be relatively so warm and moist that they will form light haze and spiral streamers or even the small cumuli known as cirro-cumulus. To a certain extent, however, it is probable that the moisture would not have risen so high without previous formation of lower ordinary cumulus, therefore, if we trace the air out of which the cirrus is formed back to its earliest history, we may find that it has once formed a part of a cumulus, or is the remainder of a cloud whose great lower part shed its rain and disappeared even the day before. Therefore the altitudes of the cirri, like the tops of the cumuli, show to what extent the air has been penetrated by the circulation due to convection from below, and how far it is being warmed and moistened in preparation for the general ascent that will soon take place in the formation of an extended storm. But there are cases when the cirri probably have an independent origin, *i. e.*, when they first form in the afternoon before cumuli are seen.

As the heights of clouds are not usually determined by reporting telegraph stations, although it could easily be arranged that they should do so, therefore all accurate observations of this datum have hitherto been made by a few special observers, and for storm studies an approximate idea of the heights of clouds must be obtained from the average results of their work; thus Dr. Vettin, from observations at Berlin, Germany, deduces the average altitudes for each month of the year about as follows:

Clouds.	Maximum height in—		Minimum height in—	
		Feet.		*Feet.*
Lower cumulus	July	5,900	February	4,800
Cumulus	do	14,000	January	9,900
Small cumuli	June July	25,800	do	21,000
Lower cirri	do	49,700	do	38,900
Upper cirri	do	85,100	do	67,200

The above figures are only relative, being based on certain assumptions, but can be applied by the observer, and in any special case can be converted into absolute data if he has determined the height of any one of these five kinds of clouds.

The fact that a cloud is but the locus of that portion of a current in the course of the ascending and descending circulation, that is favorable to the production of condensation, as has been before stated in the case of the cirro-cumulus formation, is forcibly illustrated in several special cloud formations; for example: First the case of the center of a dust-whirl occurring on warm sunny days in an otherwise cloudless atmosphere; thus in 1888, August 10, at 2.30 p. m., on the broad Pennsylvania avenue of Washington City, at Twenty-first street; wind light;

barometer high; sky blue; temperature of air in shade, 87° Fah.; dew-point, 60° Fah.; estimated temperature of asphalt pavement, 150°; a dust-whirl was seen to be forming; the diameter of the dust-bearing whirl as first noticed was about 20 feet and its height not over 30 feet; in a few seconds it had moved about 100 feet along and across the avenue northeast, but had grown rapidly upward vertically, attaining a height of about 100 feet, its lower end about 4 feet above the pavement, when suddenly, in the midst of the dusty column, was seen a narrow whitish streak or cylinder about 1 or 2 feet in diameter, reaching from 20 feet above the ground upwards 80 or more feet to the top; in about ten seconds the whole had moved very slowly horizontally, the upper portion had progressed faster than the lower, the cylinder vortex had lost its symmetrical shape, several curvatures developed in its axis, and it broke up instantaneously. In this case I conclude that the centrifugal force of the revolving vortex, which made four rotations per second in the dust-whirl a few feet outside the vapor column, had maintained a core of decidedly low pressure such that air drawn into it expanding and cooling to the dew-point gave rise temporarily to the white cloud of the cylindrical axis. From the figures above given and the laws of thermodynamic cooling, it is evident that air of the temperature 87° Fah., and dew-point 60° Fah., when suddenly elevated 5,400 feet, or when reduced in pressure by an equivalent amount, namely, from 30 inches to 25 inches, begins to deposit its moisture; hence the pressure in the center of the core must have been not greater than 25 inches of the barometer, which corresponds to the observed rotary velocity of about five turns per second. (See the formula given by Bassett, Vol. II, p. 37.)

The second illustration of this class of clouds is found in the so called "smoking" from the summits of high peaks, and may sometimes be observed to the leeward of chimneys at low altitudes. In this case the strong wind produces to the leeward of the obstacle a long series of vortices (See Fig. 52). These would be invisible were it not that the

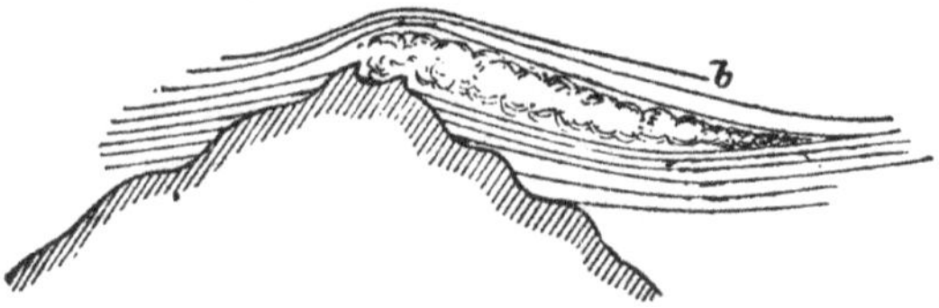

FIG. 52.

air forced up the windward side of the peak and partially cooled by expansion (as well as moistened by the evaporation from forests and snows) is now drawn into these vortices where it expands and cools into a thin delicate cloud as it passes along through the vortex, but of course becomes invisible by the re-evaporation of the moisture as soon as it reaches the end of the series of vortices at *b*.

The famous cases of "the helm-wind and helm-bar" are further illustrations of the same class of cloud. In this case (see the Report of Wm. Marriott, Quarterly Jour. Roy. Met. Soc., England, April 1889, Vol. XV, p. 105, and see also the original explanation by Espy in his Philosophy of Storms, Boston, 1841, p. 552), a moist easterly wind (see Fig. 53), flowing from Durhamshire to Westmoreland drives up the windward slope of the Crossfell range of mountains, in Westmoreland, forms a cloud capping its summit and then descends on the leeward side; in its descent the vapor evaporates and no cloud is seen, but shortly after we find that the descending air, being unable to move the mass of air in Eden Valley, rises up and passes over it; in its ascent it reaches an elevation sufficient to form a second small cloud—the "bar" that stretches up and down the valley and whose existence is evidently due to the same physical relations as that which causes the larger "helm," excepting only that there is no visible mountain for the "bar" to rest upon; the wind then passes on toward the west without forming a second bar, whence we conclude that it must move nearly horizontally but soon has risen enough to form the stratus bank in the west that is really a continuation of the stratus on the east. If the flow had continued in a series of undulatory motions there would undoubtedly have been a cloud formed at the summit of each wave, but the clouds would successively have been of smaller dimensions owing partly to the mixture of the dry-land air with the moist air from the east, and partly to the diminishing energy and extent of the undulations. Above the helm are observed the small light cumuli shown, Fig. 53, which are stationary and evidently represent the tops of standing waves high above the Crossfell range produced by the action of the latter on the east wind precisely as the rocks in the river-bed affect the motion of a stream.

A third precisely similar example occurs in the case of the so-called "Table Cloth of Table Mountain" at Cape Town (see Fig. 54, or Q. J., Vol. XV, p. 109). In this case the wind from the ocean blows up over Table Mountain; just before reaching the summit the formation of clouds begins, and continues as the wind crosses the summit until it begins its descent on the leeward side. The subsequent course of the wind is as shown by the arrows, descending and warming, then ascending and cooling, so that a second small cloud or bar is formed some distance to the leeward in mid air at *b*, but the formation of a third is not generally observed. The dissipation of cloud as it is forced down the leeward side of a mountain was well observed by me in 1878 on Pike's Peak; the account on pages 145–148 of Baker's Eight Years in Ceylon illustrates the same phenomenon on the Hackgalla Mountains, and undoubtedly it is a universal rule.

The roll cumuli that precede the advance of a cold wave, or that progress with great regularity a short distance above the ground when the air from the ocean rolls in over the land (see Fig. 55) illustrates the same feature; thus the ocean breeze at *o* on striking the shore curves up as at *b* and proceeds thence to *c c* in a succession of undulations. The observer on the ground below feels a little of this by virtue of the oscillations of the force and direction of the wind recurring every few minutes. The observer at *a* can easily detect the presence of the up-rising currents overhead by means of an ordinary kite, as I myself did at Ocean Beach, on the New Jersey coast, in the summer of 1876.

The careful observation of the kite showed that the ocean breeze was blowing overhead before it was felt at any point on the shore, and as the morning progressed its boundary line *o n b* moved landward, so that when felt on the land, as at *p*, its boundary was *p p*, and an observer a short distance back from the beach, while not feeling the full breeze near the surface of the ground, could by means of the kite recognize its

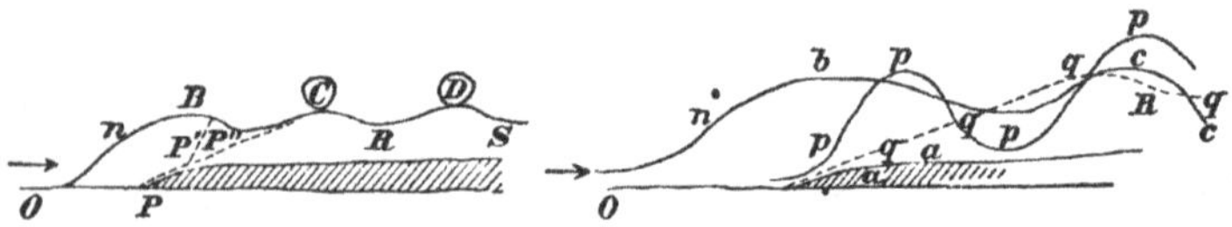

FIG. 55.

presence a short distance above him; probably the early morning boundary *o n b* became changed during the day until, when the breeze was at its maximum, the boundary was *q q q*. When a breeze of this character is laden with moisture the portions *b c*, *p p*, *q q*, become visible as roll clouds, or cumulo-stratus, or the undulating bottom of a continuous cloud.

Such rolls or whirlwinds with horizontal axes are described by Mohorovicie in the Met Zeit., 1889, page 56, over the Bay of Buccari where the "Bora" flowing over terraced or undulating ground is thrown into cylindrical waves, which combined with the forward motion give a wave or cylindrical surface, as in Fig. 55.

Such roll clouds attend the advancing fronts of northers, blizzards, thunder-storms, Pamperos, Solanos, and other winds of this class. The descending heavy air gives us the destructive gust; the ascension is so gentle that we overlook it except in the case of tornadoes. The average angle of inclination of winds to the ground, in European works on windmills, is said to be 17°, but this relates only to the severe gusts. In May, 1856, in New York City, and in June, 1858, in Lansing, Mich., on each occasion about 6 a. m. of a clear morning, I observed such delicate roll clouds, of only a little elevation, rapidly move eastward as the cool air to the west flowed down and rolled up the moist warm air on which the rising sun had begun to have an effect.

The relative temperature of the air at the surface of the ground decides the question as to whether a certain mass of air shall rise by its buoyancy, or whether a certain other mass of air shall by its density underrun the lighter. But as we before said, neither of these processes constitutes a cyclonic storm; in the latter there is a decided up draft due to the buoyancy of the stratum of air at the level of the clouds and not merely to the relative buoyancy of that about to rise up; it is granted that there would be no ascent without a buoyant surface stratum, but having ascended there would be no further up draft without a buoyant cloud stratum. The formation of a cloud and inception of a storm may depend upon surface conditions, but the growth of a storm must depend primarily on the cloud conditions, and secondarily on the condition of the surface air that is drawn up to feed the cloud; therefore, in studying the progress of the storm our first question is as to the relative densities of air at the cloud level on all sides of the storm. By means of a table of densities we are able to determine the lines of equal density for the surface stratum or rather for an ideal sea-level stratum. It is now important to do the same thing for a higher level for which we shall assume an altitude of 5,000 feet. In order to ascertain the specific gravity of the air at this level we need to know its temperature, pressure, moisture, and the force of gravity. These may be obtained from the surface conditions with a considerable degree of approximation if we take account of all attending circumstances.

(*a*) *Vapor.*—If clouds are visible, or rain falling, so that we are justified in assuming that up to the cloud limit the air above us is like that which was at the earth's surface a few hours ago at some point to the windward of our station, we may go back to that region and from the dew-point there reason to what the dew-point must be at the cloud limit above us or within the clouds themselves.

If no clouds are visible or for regions above the clouds we must assume that the distribution of vapor is according to Hann's formula:

$e = E\, 10^{\frac{-h}{6517}}$ where e and E are in millimeters and h in meters,

or

$e = E\, 10^{\frac{-h}{21310}}$ where e and E are in inches and h in feet.

The vapor tension is independent of the vapor that exists as fog and rain in the air, but the effect of the weight of the moisture upon the barometer at the bottom of the atmosphere is, on the contrary, just the same whether it exists as vapor or as fog. This remarkable rule is too frequently overlooked or misunderstood, and results from the fact that the falling fog quickly acquires an impaired velocity such that the resistance balances the acceleration due to gravity and the whole effect of gravity is added to the air. Thus a given volume of ordinary atmosphere has the same mass and the same density and the same specific gravity in whatever state its included moisture may be.

(*b*) *Temperature.*—We need the average temperature of the lower 5,000 feet of air at each point of observation. This temperature we adopt from the following considerations based on the variations of temperature with the winds and weather and the time.

For the clear air in the rear of the storm a uniform rate of diminution of temperature with height may be adopted, at least, for those regions in which the wind is strong enough to assure us that the air at and below the cloud layer is being well mixed, in which case the rate of diminution would be about one-half of that due to adiabatic changes, or 0.6° C. for 100 meters, or 0.3° Fah. per 100 feet or 10° to 3,000 feet, but in the cloudy portion of the area, where we have every assurance that the clouds are formed of air that was at the surface of the earth a few hours before, we may assume that the temperature of the lowest part of the cloud is nearly the same as that of the dew-point observed in the air at that surface a short time before. The temperature will really be somewhat less than that of the dew-point, and as a rough approximation it may be allowed to assume temperatures at the elevation of base of cloud that are lower than the dew-point by one-fifth of the depression of the dew-point below the temperature of the air

$$t = DP - \tfrac{1}{5}\ (T - DP).$$

More refined theoretical relations could be given, but this degree of approximation will suffice for the present. Above this level within the cloud we adopt Hann's table of adiabatic cooling.

(*c*) To ascertain the pressure of the adopted upper level of 5,000 feet, some one of the many hygrometric formulæ must be employed and the selection must depend upon the accuracy with which we are supposed to know the average temperature and vapor tension of the stratum of air in question, and this will vary principally with the character of the sky as to cloudiness and haze.

The formulæ themselves that one naturally thinks of are the older ones of Laplace, Plantamour, Gauss, and Babinet—all given in Guyot's Smithsonian tables—or the newer ones of Ruhlmann (1870), Schreiber (1877), Hann (Vienna Sitzungsbericht, 1876, LXXIV), Upton (An. Rep. C. S. O., 1882, pp. 830–837), Ferrel (Met. Researches, Part III, Washington, 1882, and Recent Advances, 1885, p. 396), Sprung (Met. Zeit. 1888, V, p. 460), Koppen (Met. Zeit., 1888, V, pp. 369 and 470), and for many purposes the latter, which neglects the moisture, will suffice, but it is best to take at least approximate account of this component of the atmosphere, and this I do in the following formula, which is modified from that of Upton:

$$\log. b = \log. B - \frac{h \text{ (in meters)}}{18442.5\,(1+0.0036774\ T+0.189\,\frac{E}{B}\left(1+10^{\frac{-h}{10091}}\right)(1+0.0026319 \cos 2\,\varphi)}$$

Where B is the observed lower pressure, T and E the adopted average temperature and vapor tension of the whole column and h the adopted altitude from which the pressure b is derived.

The vapor tension can be tabulated for a given h, E, and B, and this term can be treated as a correction to the term 0.0036774 T; so that the equation may be written—

$$\log. b = \log. B - \frac{h}{18442.5(1+0.0026319 \cos \varphi)} \times \frac{1}{1+0.0036774\left[T+\frac{0.189(1+10^{\frac{-h}{10091}} \cdot \frac{E}{B}}{0.0036774}\right]}$$

This last fraction may therefore be tabulated for value of T and $\frac{E}{B}$ or T and ΔT.

The logarithm of the density at 5,000 feet above any point can now be computed from the Landolt and Bornstein table, giving

$$\delta = \frac{.0001293052}{1+0.003670t} \, \frac{b-\frac{3}{8}e}{760}$$

as for the density at low level and we can draw lines of equal density, which we will call isostathmic (isostathmos or equally balanced) lines or isostaths.

The isostath that passes through the center of a whirl divides the lighter air from the heavier or the air that must go up from that which must go down in the gradual interchange due to buoyancy; but the air that must go up forms the cloud and rain and the new region of greater buoyancy, therefore the lower air will tend to move toward the side of the lighter density unless there is some other reason or cause pushing the heavier air upwards; the only other causes to bring about this latter result are the rotation of the earth and the topography of the ground relative to the direction of the wind. The latter is an important factor in storm motions and our problem now is to determine on which side of the storm center the greatest quantity of cloud will be formed in consequence of the elevation of the lower air by the two causes internal buoyancy and orographic gradient. When the tendency of the air at any time and place to move toward a given direction is thus determined we shall have simply to graphically represent this tendency and integrate the resulting equations of motion or their graphic presentations in order to obtain the resultant movement of the storm center.

The most expert observers, and especially the shepherds, sailors, and farmers depend much on the appearance of the clouds, and it would be natural, in making extensive weather predictions for distant regions, when the distant sky is hidden, as it is, from our personal observations, to require all possible telegraphic information as to the clouds. The elements of the system of nomenclature, introduced by Luke Howard early in the century, afford us the only names practicable to introduce for the use of ordinary observers; the necessity of having such information about clouds became apparent in my predictions of 1869 and 1870 in Cincinnati, and a system of telegraphic cipher dispatch was then prepared that allowed of economically sending the following items of information: (1), the kind as shown by Howard's terms; (2), the amount or proportion of sky apparently covered by clouds; (3), the direction and rate of motion; and these items were given separately for each of the several layers when such were present. This system of cipher dispatch was introduced into the use of the Signal Service in the summer of 1871 in place of the numerical cipher previously used, and a tri-daily cloud map was then begun, showing graphically for the whole country the kind, amount, and direction of each layer of cloud.

The study of these maps in a few weeks enabled me to announce the general rule (entirely in accordance with Espy's, Redfield, and Ferrel's theories) that the lower clouds moved toward points to the right hand of the surface winds; the upper-cloud directions trend to the right of the lower clouds; or, in general, the higher we ascend the more the movement of the air deviates to the right. These results were further confirmed by a study of the balloon voyages of Prof. S. A. King. (See the bulletin of the Philosophical Society, of Washington, Vol. I, pp. 36 to 38, where I have so arranged the winds experienced by King in his ascensions as to show that the courses at great heights are from 90 to 135 degrees to the right of those below, and that the average velocity of the upper currents shown by seven balloon ascensions in different parts of the country on July 4, 1871 and 1872, was about four times that of the surface winds, as shown by Signal Service anemometer observations.) A similar law as to the relative direction of upper and lower currents was soon afterward published by Rev. Clement Ley, in his Laws of the Winds, and is, in fact, easily deducible from the general law of cyclonic and anti-cyclonic movements in the northern hemisphere. By means of the general law of deflection to the right we can approximate to the direction of upper currents even when no clouds are present or when lower clouds hide the upper.

CHAPTER VI.

CLOUDS AND BUOYANCY.

1. It is evident from what precedes that the winds at the surface of the earth are the inflow due to the uprising of buoyant air, and the character of the winds depends upon the thermal relations of the ground, the air, the clouds, and the sun.

If it can be shown that the lower or upper strata will be heated and rise without formation of rain, it follows that there will be a reversible process, and no great disturbance will arise. A larger disturbance is started when clouds are formed and radiate their heat. If merely clouds are formed the loss of heat from their surface by radiation may be either less than, equal, or greater than, the solar heat absorbed, depending upon the shape of the cloud and the properties of cloud particles as to absorption and radiation. If less, the cloud must increase in buoyancy, and grow in size or alter its shape, until the heat radiated from the surface equals that absorbed by the surface from solar or terrestrial radiation; but the growth of the cloud will be greatest on the sunny side, so as to produce larger shadows on the ground beyond. The sunny side will rise to higher elevations, the radiation from the elevated portions becomes greater; the radiation from the shaded portions, plus the extra radiation from the higher portions, balances the absorption of heat on the illuminated portions; the maximum size of the cloud possible under the given conditions of temperature, moisture, and sunshine, is maintained until the gradual diminution of solar radiation as the sun declines in the west makes the loss of heat by radiation greater than its gain by absorption when the cloud begins to diminish in size, unless the internal heat added to it by a superabundance of condensing vapor brought up by the earth below prolongs its growth for a few hours. In this case the heat thus added may, during the evening, partly replace the heat added by solar radiation during the day-time; but this supply of heat rapidly ceases during the evening hours, since the lower strata near the earth have themselves cooled and can not be drawn up into the cloud, therefore, eventually, during the night, the cloud loses both sources of heat, namely, solar and terrestrial radiations, and the latent heat of condensing vapor, and it dwindles away under the influence of loss of heat by radiation from its own surface. The radiating power of the cloud per unit surface must be at least one-eighth times the absorbing power for solar heat in order to maintain thermal equilibriums.

It is evident, therefore, that up to this time the cloud has been merely a perpetually renewed appearance and that the vapor particles which constitute it are in a state of perpetual change.

The particles are rising upwards within the cloud and again descending near the outside of the cloud; while rising they are increasing in size, while descending they are diminishing in size; if in their ascent they become very heavy they may eventually drop down nearly vertically, be in part evaporated, and reach the ground as rain; if, however, they remain small while ascending, and are carried out to the outer descending whirls that form the surface of the cloud, they will be wholly evaporated and disappear from sight; in this way the outer limit of the cloud is defined.

2. Eventually the re-evaporation of cloud particles consumes the latent heat that had been liberated by their condensation, therefore, the only heat that remains in the cloud as the permanent motive power by which to overcome all forms of resistance, such as viscosity, fluid friction, slip on the earth's surface, and resistance to impact, and to do the work of lifting up heavy air, as at night time, when the buoyancy of the surface strata will not cause them to rise of themselves,—the only permanent source of power is, I say, the heat left in the cloud from the vapor that condensed and fell as heavy drops of rain without being re-evaporated in its fall. So long as rain-fall continues a corresponding amount of heat or power or energy is left in the cloud to continue its growth in spite of all obstacles. Now the rain comes mostly from the lower air just freshly drawn up to the clouds; the clouds act as the upward-suction power and after the rain has fallen the cloud receives and makes use of the evolved latent heat.

The obstacles to continued growth of cloud and development of rising currents are: first, the intermixture of cool dry air with the moist warm air that constitutes the vortices; second, the radiation directly into space and to a very slight extent into the surrounding atmosphere from the warm air and vapor at the surface of the cloud. The quantity of heat corresponding to the rain-fall in one minute, diminished by the heat radiated from the same cloud in that minute, and increased by the solar or other heat absorbed by that cloud during that minute, represents the momentary rate of growth, or rather the power to grow.

3. Experience shows that columns cease to rise from the ground to form new cumuli after 3 p. m., but that the growth of those that had begun before, continues, and that these now grow by the force of an internal buoyancy; that is to say that the layers of air near the earth surface which had in the morning been so much overheated as to rise up several thousand feet, are now relatively cooler than the upper layers; the insolation is no longer sufficient to raise the vertical temperature gradient enough to send them up into the warmer layers above so rapidly as they rose in the morning; on the other hand, the

warmer air just above the lower layer still has cooler air above it, and will continue its buoyant ascent as in Fig. 56, therefore, the cumuli after 3 p. m. generally grow by making drafts upon the warmer air at considerable elevation above the ground but below their own bases, therefore, as the evening progresses, this source of supply must be at successively higher elevations, but the base of the cloud is at a minimum elevation at 4 p. m., after which it retires upward; so far as this supply of warm moist air from below is concerned the top of the cloud will be at its maximum elevation when the base is at its minimum. The stratum of convective activity rises as shown also in Fig. 56; after 5 p. m. the top of the cloud receives from the direct sun more heat than the bottom does from the ground, and is specially apt to grow by enlargement and overflow at top, giving rise to the overlaying glistening white summit crest S, Fig. 56.

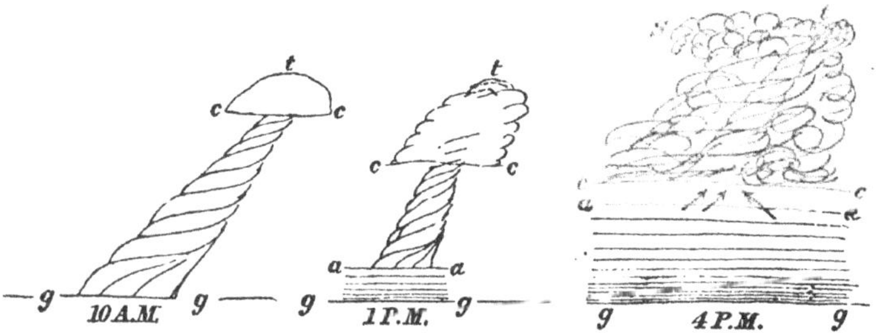

FIG. 56.

The relation between shade and sunshine at the earth's surface, as given by the sunshine recorder, which is really a relative duration, can be converted into relative quantities of heat, received at the earth's surface, and at the cloud surface, respectively; thus if during any hour the average altitude of the sun is A degrees and the corresponding amount of heat received by a unit of horizontal surface is H, but by a unit of normal surface is N, then during X minutes of sunshine at the earth's surface a unit surface receives HX calories, and during Y minutes of shade the cloud surface receives NY calories on the assumption that the cloud surface is normal to the sunshine, as is frequently the case on the sunny side of the great cumulus clouds. As a general average for large regions of the earth it would be better to adopt a method similar to that used by Zöllner in computing the brightness of the surface of the moon, which is a case quite parallel to that of sunshine on the tops of clouds; but the following diagrams sufficiently illustrate our present idea. Fig. 57, which gives us by curves the relative amount of heat received at the surface of the earth as compared with that absorbed by the atmosphere and clouds, or the ratio

$$\frac{\text{insolation of ground}}{\text{absorption of air and cloud}}$$

at each moment from sunrise to sunset, first, for perfectly clear weather, second, for partly cloudy weather. The third curve gives us corresponding absolute absorptions at the surface of the clouds. The fourth curve, or the difference between these, gives us the datum that determines the relative frequency and quantity of cumulus clouds, showing their rapid decline towards sunset.

4. In a dry atmosphere as much air descends as ascends, and the descending is warmed as much as the ascending is cooled, so that by mere vertical interchange, the atmosphere as a whole loses no heat. Similarly in a moist air, which by ascending cools and forms clouds and by descending evaporates those clouds; the atmosphere as a whole loses no heat by this operation alone. There is in both these cases only a temporary transfer of heat to an upper region. The result would be an accumulation of heat in the higher portions of the atmosphere, or in general in the colder portions since by similar horizontal convection a transfer of heat eventually takes place from the equatorial to the polar regions. It is only the actual loss of heat that takes place by radiation that prevents the progressive warming of the atmosphere and explains the average uniformity of temperature that prevails from age to age.

The progress upward of the surfaces of equal heat and moisture by vertical convection is paralleled by the progress northward of the isotherms or the surfaces of equal heat and moisture as determined by the horizontal convection. As the lower air requires time in order to get rid of its excess of heat by vertical convection, so also time is required in the horizontal convection; as the lower air at lower latitudes, owing to the high sun and feeble winds, retains more solar heat from day to day than it can lose by radiation and vertical convection, it grows warmer until larger convective processes carry the heated air up to higher altitudes and polar latitudes.

This diurnal and annual periodic delay brings our maximum temperature at 2 or 3 o'clock p. m. and in August or September instead of at noon and in June, respectively; it also causes any periodicity in solar radiation to distribute itself gradually over the earth so far as concerns its effect on terrestrial temperature in a manner similar to the distribution of vapor from the Krakatoa eruption, which latter required a year to reach the northern limit of the temperate zone. The local and general motions in the atmospheric currents are, therefore, in these respects parallel to those of the ocean. Periodic phenomena, manifested first in the tropical regions and thence propagated polewards, find the lengths of their periods increasing and the amplitudes diminishing with time or with their progress over the surface of the earth according to a law expressed by exponential co-efficients such as those introduced very frequently by the integrations of the formulæ for wave or periodic motions in viscous fluids, and such is the analogy between the mathematical features of the two problems that the results obtained for viscous fluids may be directly interpreted into homologous results for motions on the earth's surface.

5. It is the action of gravity that imparts to dense or rare air the quality of weight that we distinguish by the words heavy or light, respectively, and in considering the buoyancy of a quantity of gas it is necessary to keep in mind the distinction between density and weight. The relative density, or specific density, is a question of mass, but the relative weight, or specific weight, is a question of the action of gravity upon that mass; the buoyancy of one gas with respect to another is the difference of the weight of the two when under the action of the same force of gravity. If, however, the two masses are not at the same spot, *i. e.*, under the influence of the same gravity, then the buoyancy will not cause one to rise above the other unless they are both on the same level surface, in which case the heavier one pushes the lighter one aside. The total displacing power, whether it produces ascension of the lighter gas and is called buoyancy, or whether it produces horizontal flow of the denser gas, or as is generally the case, both of these combined, is the difference in weight of equal volumes of the heavier as compared with the lighter gas; this difference in weight or the buoyancy depends on the densities and the absolute forces of gravity acting on the two masses, so that effect of the variation of gravity with altitude and latitude may be appreciable. In order to make pressures measured by the mercurial barometer comparable among themselves a correction for variation of gravity must be introduced as affecting the height of the mercurial column. In order to ascertain the relative weight of the same mass of air under different forces of gravity, there must be introduced a corresponding correction for the action of gravity upon that mass of air. This consideration leads us to the formulæ already given for density and pressure, and to the following additional tables:

TABLE I.—*Weight of a cubic foot of air at a pressure of 30 inches for various temperatures and moistures.*

Temperature.	Depression of dew-point.							
t	0	5	10	15	20	25	30	∞
0	602.77	....	....	....	....	....	....	603.21
10	589.40	....	....	....	....	....	....	590.04
20	576.54	....	....	....	....	....	....	577.44
30	564.08	....	....	....	....	....	....	565.35
40	552.00	552.5	553.0	....	....	....	....	553.77
50	540.21	540.9	541.5	....	....	....	....	542.65
60	528.62	529.4	530.1	530.6	531.0	....	....	531.97
70	517.17	518.1	518.9	519.6	520.0	520.4	....	521.70
80	505.74	507.0	508.1	508.9	509.5	510.1	....	511.82
90	494.28	495.4	496.4	497.3	498.0	498.7	499.3	502.32
100	482.72	....	....	....	....	....	....	493.18

In Table II are given the buoyancy of a cubic foot of saturated air under a pressure of 30 inches, both in absolute measure and relative to its own weight. We see from this that through the influence of moisture alone a buoyant pressure is exerted upwards that may amount to more than 1 per cent. of the total weight of the moving mass. The buoyancy imparted by heat alone is shown by comparing the numbers given in the first column and the ninth column of the preceding table for different temperatures, and the results are given in the fourth to the seventh columns of Table II, which show that a rise of 10 degrees in temperature affects saturated air slightly more than dry air. The absolute effects are given in the columns of weights and the relative effects in the columns of ratios.

TABLE II.—*Buoyancy due to moisture and temperature.*

Temperature.	Maximum buoyancy due to moisture.		Maximum buoyancy due to heat for a rise of 10° Fah			
			Saturated air.		Dry air.	
	Excess of weight of dry over saturated air.	Ratio of excess to total weight.	Excess of cooler over warmer.	Ratio of excess to total weight.	Excess of weight of cooler over warmer.	Ratio of excess to total weight
	Grains.		*Grains.*		*Grains.*	
0	0.44	0.001	....	....	....	....
10	0.64	.001	13.37	0.023	13.17	0.022
20	0.90	.001	12.86	.023	12.60	.021
30	1.27	.002	12.46	.022	12.09	.021
40	1.77	.003	12.08	.022	11.58	.020
50	2.44	.004	11.79	.022	11.12	.020
60	3.35	.006	11.59	.022	10.68	.020
70	4.53	.008	11.45	.022	10.27	.019
80	6.08	.011	11.43	.022	9.88	.019
90	8.04	.016	11.46	.023	9.50	.019

The formula for the mass or absolute density of air (*i. e.*, using water as the standard) at a temperature of t, vapor tension e, and pressure p, (both e and p and p_0 have already been given and are to be expressed in the same units) is as follows:

$$\delta_a = \frac{0.001293052}{1+0.003670\,t} \cdot \frac{b-\frac{3}{8}e}{760}$$

Where b and e are expressed in barometric heights corrected to standard gravity. The specific weight Γ or the action of gravity upon the mass whose density is δ is given by the formula

$$\Gamma = G\,\delta = G_0\,\delta\,(1-0.00259 \cos 2\,\varphi)$$

where G expresses the weight of a unit mass under local gravity and G_0 its weight under standard gravity (at 45° latitude and sea-level) and φ is the latitude. The absolute values of the forces in action are given by considering the inertia or energy that could be given in a unit time to the unit mass by the motive action of a unit force of gravity: Call this unit force f; then a unit volume of air, whose mass is δ, will in its unit time, one second, receive from the actual force of gravity an acceleration of g units of velocity or an increase of $f \times \delta \times g$ units of energy.

In absolute units we have:

$\delta = 1.29305$ kilograms per cubic meter as the absolute density of dry air at 0° C. expressed in the units of mass; Γ, or the specific weight of dry air, is 1.29305 G; $g = g_0 (1 - 0.00259 \cos 2\varphi)$ = the local acceleration due to gravity or the local force of gravity, where g_0 expresses the standard value of the acceleration of gravity at 45° latitude and sea-level and is approximately 9.8065 meters per second or 9.8065 units of velocity; this is really apparent gravity, namely, the attraction of the earth plus the vertical component of centrifugal force. The latter effect should be restored to the apparent gravity if we would obtain the total gravitation by which the earth acts upon bodies that are not rotating with it. For motions in the atmosphere the differential motion of the air with regard to the earth has so little effect that the value of apparent gravity is ordinarily used.

The mass of aqueous vapor in kilograms per cubic meter of saturated atmosphere at temperature t is:

$$\delta_v = 0.622 \quad \frac{1.293052}{1 + 0.003670\, t} \cdot \frac{e_t}{760}$$

For convenience of memory it may be noted that the number that gives the vapor tension in millimeters of mercury is very nearly the same as the number that gives the weight of the vapor in grams per cubic meter.

Dew-point.	Mass per cubic meter.	Dew-point.	Mass per cubic foot.
°C.	*Kilograms.*	°Fah.	*Grains.*
−20	0.001042	0	0.545
−10	.002283	10	.841
0	.004869	20	1.298
+10	.000057	30	1.969
20	.017148	40	2.862
30	.030079	50	4.089
40	.050674	60	5.756
		70	7.992
		80	10.949
		90	14.810
		100	19.790

6. The condensation within a cloud leads to the deposition of some, namely, a certain percentage of its moisture, which, by its own weight, settles down out of the cloud and, if it does not reach the earth as rain, is by evaporation diffused through the lower atmosphere, and in either case is lost to the cloud itself.

The average value of the quantity of rain that falls on any occasion, taking all the rains together, is found, for instance, for Washington for the year 1888, to be:

$$\frac{\text{Annual quantity of rain}}{\text{total number of rainy days}}=\frac{31}{133}$$

or about one-fourth of an inch; but for storms of any extent or severity, the average over large areas is considerably more than this, and may be assumed to be 1 inch in depth in twenty-four hours, diminishing from nothing at the circumference of the storm up to a maximum of 4 inches near the center of the severest storms, and even to 10 or 15 inches in phenomenal cases. It is important to have an approximate connection between the quantity of moisture that falls, the total quantity in the clouds, and the total quantity in the atmosphere, and this is approximately accomplished by the use of the formula given by Hann, according to whom the total mass of vapor in a column of atmosphere 1 meter square and expressed in kilograms is

$$\text{(I)} \qquad Q=\left(0.00106\frac{E}{1+at}\right)\times 2830\left[1-10^{\frac{-h}{6517}}\right]$$

where E_0 is the tension of the vapor and t the temperature at sea-level, and h the height to which we wish to extend our computations. For the whole atmosphere $h=6517$ and we have simply

$$[Q]=2830\times\left(0.00106\frac{E}{1+at}\right)$$

or 2830 × the mass of vapor in a cubic meter of air at the earth's surface.

For less heights the factor given by Hann is as follows:

[See Z. O. G. M. 1884.]

Height.	Factor
Meters.	
1,000	842
2,000	1,433
3,000	1,849
4,000	2,141
5,000	2,346
6,000	2,491
7,000	2,591
8,000	2,664
9,000	2,711

The preceding gives the weight in kilograms of the moisture contained in a column 1 meter square, extending to the top of the atmosphere. This water, if all lying upon the ground, would cover 1 square meter to the depth of one ten-thousandth part of the quantity given in formula I, or

$$H=\frac{[Q]}{10000}$$

so that we may write H in meters of depth $=0.2830\times$ by the weight in kilograms of the water in 1 cubic meter at sea-level.

This, in conjunction with the previous table, gives us from 0.001 to 0.007 meters, or 10 to 20 inches, as the equivalent depth of the layer of water. Now, within a given storm area, although the currents of air that rise over one region may deposit their moisture over another, yet by taking a general view we may assert that over the whole region there is an average rain-fall and an average quantity of atmospheric vapor. The latter for extensive storms in the temperate regions averages about ten times the former, so that the percentage of vapor that falls as rain out of any given cubic meter of air varies from 10 per cent. down to nothing, and will average not over 5 per cent. for the atmosphere as a whole. The excessive local rain-falls are due not to the fact that nearly all the rain is condensed out of the atmosphere, but to the fact that fresh air is rapidly taking the place of that which has just before given up some of its moisture, and in its turn gives up its own percentage of moisture. This implies that there is a rapid convection going on overhead, precisely such as we see in every storm, and other things being equal, the more severe the precipitation at any spot the more rapid must have been the convection over that spot; moreover, as the upward convection is one that is most efficient in cooling, condensing, and the formation of rain, we infer that, other things being equal, the heavier downfall implies a stronger upper current above, and this latter as we have seen before is determined principally by the orographic gradient in the immediate neighborhood.

The exact amount of rain-fall corresponding to a given percentage of moisture condensed out of the atmosphere is given most conveniently by the diagram devised by Dr. Hertz, and reproduced in Fig. 58, from Met. Zeit., 1884, Bd. I, p. 421, and pl. 7. On this diagram will be found several systems of intersecting lines that graphically present the adiabatic change of pressure, namely:

(1) The horizontal lines represent the temperature of the air.

(2) The vertical lines represent the pressure.

(3) A system of diagonal lines ascending toward the right, which is called the "alpha" system, and represents the progressive change in temperature and pressure of a mass of air as it rises in the atmosphere, the act of rising being indicated by a motion toward the left, namely, down the "alpha" lines.

(4) Another system of inclined lines, which is designated as the "gamma" system, and represents the progressive change in temperature and pressure after the ascending air has formed a cloud, but is still carrying all of its moisture up with it. The gamma lines are suddenly broken when they reach the line of zero degrees or freezing point, because at that point the formation of ice causes an evolution of heat without any sudden change in the altitude of the ascending mass.

(5) Another system of lines is given on the diagram, namely, the "beta" system. These are dotted lines and are marked on the left-

hand side with figures to show the weight of moisture accompanying a kilogram or unit mass of air.

To illustrate the use of this diagram let the unit mass of air start at a temperature of 27° C. at a point whose pressure is 750 millimeters; we find this point at the upper right-hand corner of our diagram, and if we wish to know how high it is above sea-level we drop from that point vertically downward into the subsidiary table at the bottom of the page, where we find the altitude to be about 100 meters above the sea.

If now this air be carried upwards to lower pressures, expanding and cooling as it rises, we find its condition at any moment by following along the dash and dot line, as given on the table, through the initial point and parallel to the nearest alpha line. When the mass has attained the altitude of 640 millimeters of pressure the indicator point on the dash and dot line cuts the temperature line for 13° C.—this shows that the temperature has been reduced to 13 degrees. Let us suppose this to have been the original dew-point of the air, then, at this elevation, a haze will begin to form; we, therefore, find the further temperatures and pressures that will attend its further ascent by causing the dot and dash lines to make a turn at this point and become parallel to the nearest line of the gamma system. The indicator follows this line down until the temperature has cooled to 0°, which occurs when the pressure is about 482 millimeters; then sufficient latent heat is evolved by the freezing of the fog particles contained in the air to keep the mass at a constant temperature of 0° until it has risen to a level, where the pressure is about 461 millimeters. After this as the mass rises the temperature diminishes and the dash and dot line follows the new gamma curve representing the further ascent of the air. All this is on the assumption that no moisture is lost from the original mass of moist air, and this is approximately true within a large cloud until a very considerable elevation has been attained.

The quantity of moisture existing in the air at any point in its ascent may be divided into two portions, the vaporous and the liquid; the sum of these two is constant so long as none falls down, and is, therefore, equal to the original quantity with which the air started in its ascent. This total amount is expressed graphically by means of the figures on the beta lines: thus, at the point where the dash and dot line makes its first bend, and where condensation begins to occur; the corresponding beta line would indicate about 12 grams of moisture to the cubic meter. As we follow this dash and dot line further down into regions where, because of its coldness, the air can not retain all this moisture as vapor, we can ascertain how much vaporous moisture exists in the air at any point by drawing through it a short beta line and reading off its distance between the two nearest standard beta lines given on the chart: thus, at a pressure of 605 millimeters, and a temperature of 11° C., our cubic meter contains 10 grams of vapor; at a temperature of 0° and pressure of 473 millimeters, it contains about 6 grams; at a temperature of −20 and a pressure of 306, it contains a little less than 2 grams. The differences between the numbers thus obtained from the scale readings on the chart and the original vapor contents of the air gives the quantity that must be contained in the original mass in the shape of cloud, rain, or snow, and this, in the present illustration, as shown by the above numbers, is 1, 5, and 9 grams respectively; that is to say, at the altitude corresponding to 605 millimeters of pressure, there are 10 grams of vapor and 1 gram of cloud particles; at 473 millimeters and 0° C., the air contains 6 grams of vapor and 5 grams of rain, but at 462 millimeters and 0° C., it contains 6 grams of vapor and 5 grams of snow or ice; finally, at 306 millimeters, it contains 2 grams of vapor and 9 grams of snow or ice.

It would follow from the above, if there were no precipitation toward the ground, that the top of a cumulus cloud should contain a much greater quantity of rain or snow particles per unit volume than the lower portions, but the contrary is probably the case; that is to say, one-half or more of the 9 grams in our illustration settle down to the lower part of the cloud, but so large a portion as this does not reach the earth's surface, at least not directly under the location at which it starts to form. I estimate the total proportion that reaches the ground from a cloud during the whole of its ascending course as not over 5 per cent. for cumulus clouds or ordinary thunder-storms or tornadoes, and not over 10 per cent for extended cyclonic storms, where the ascending gradient is very gentle. Let us assume for illustration that as much as 10 per cent. of the liquid or solid moisture falls from the clouds as fast as it is condensed after the air rises above 640 millimeters of pressure, and consider the effect of this on the dash and dot line indicating the temperature of the ascending cloud. By the loss of this vapor the cooling process ceases to be strictly adiabatic. A small quantity of heat is continually being abstracted as the specific heat of the falling moisture; the specific heat of the small quantity of liquid thus lost is a very small quantity compared with the specific heat of the remaining 90 per cent. liquid and the latent heat of that which remains as vapor and the specific heat of the large quantity of gas that always remains. The effect of this loss of heat is to diminish the potential temperature, as that term is used by Bezold, and, therefore, the buoyant power of the remainder for any given pressure or altitude. We can estimate the effect numerically by the following process:

Let the original dot and dash line for the dry stage of the rising air be imagined continued downward in our diagram parallel to the original alpha line, until it intersects any given pressure line such as that for 462 millimeters which it intersects at about a temperature of − 15°; this is then the temperature that the original mass of air would have attained if it had cooled without the evolution of latent heat; the actual evolution has, however, had the effect of giving the air a temperature of 0° at this pressure as shown by the previous figures, in other words, the evolution of latent heat has raised the temperature of the mass 13 degrees above what it otherwise would have been; and this was due to the conversion into snow of only 5 grams out of the original 11 it contained; now, if 10 per cent. of these 5 grams had already been lost as rain or snow, the effect of this loss upon the 13° of increase of temperature would approximately be given by the expression 13° × specific heat of one-tenth of 5 grams ÷ latent heat of 5 grams of water + latent heat of liquefaction of 5 grams of snow + the specific heat of 1 killogram of saturated air. This, therefore, gives us an almost inappreciable quantity showing that even for the heaviest storms we shall, so far as this source of discrepancy is concerned, obtain the temperature at any altitude in a cloud with sufficient accuracy by the use of Hertz's diagram, based on adiabatic considerations.

The total heat thus contributed to the atmosphere, after such precipitation of rain as occurs in a storm, is but a small per cent. of that which remains latent within the remaining vapor molecules. In a general way we may say that the precipitation throughout the whole world must year by year balance the evaporation, consequently the sum total of the heat left in the atmosphere by falling rain and snow must equal that consumed in its evaporation. If on the average 10 per cent. of its moisture falls to the ground in each storm and 10 per cent. of its latent heat is left in the air, then ten such storms would be needed to bring about the thermal balance between that mass of air wherever it may have gone to in its travels over the earth and the original quantities of heat consumed in the evaporation of its moisture. Now, practically, the evaporation is greater in the tropical regions than the rainfall, and in general we may divide the world up into regions where evaporation is greater and where it is less, respectively, than the rainfall; the latter regions are polar and continental and owe a great deal of their warmth to the fact that the precipitated rain of tropical and oceanic areas falls upon them.

Within a given storm the descending currents warming by compression evaporate any liquid particles they may have, and the heat thus consumed balances that which the same current, when rising in some previous part of its history, lost by cooling or gained by evolution of latent heat; there would, therefore, be no motor power to maintain the existence of such currents as against the steady resistances of the earth's surface, but this motor is now partially supplied in the small percentage of latent heat that is left in the cloud by the moisture that falls from it. This latent heat is, in the present illustration, one-tenth of the latent heat evolved by the 5 grams that were condensed into snow; its effect upon the temperature of the mass is one-tenth of the 13 degrees, and in general its contribution toward the permanent maintenance of the storm is one-tenth of all the thermal forces at work in the storm. It is evident that the maintaining power of the storm is greater when snow or ice is formed and actually falls as such to the ground than when rain only is formed; thus, in our present case if the air attains to a height whose pressure is 306 millimeters, thereby precipitating 9 out of its 11 grams as snow, and then loses 10 per cent. of that by its fall to the earth, the latent heat left behind corresponds to twice that that was calculated for the elevation 462 and is ten times that for 605 millimeters.

7. But the buoyancy of a cloud is affected largely by another consideration, namely, its absorption of the radiation to it from the sun and the earth. The temperature of any mass (A) above the surface of the earth or above the upper surface of the cloud is the result of several factors: it is due, first, to the difference between the loss by conduction and radiation upwards of the heat existing in the atmosphere at A and the gain due to the heat received at A by radiation from the earth and the atmosphere and the sun; second, to the difference between the heat convected into A by the motions of the atmosphere and the heat similarly convected away from A. The general equations of motions in fluids under the influence of changes in heat were first given by Fourier in a fragmentary essay published after his death in 1834 (Mem. l'Institut), and his equations would apply to the history of the growth and decay of a storm, but the analysis is so difficult that mathematicians have preferred thus far to spend their strength upon the "steady motions," the mathematical study of which began in 1806 with the paper by Lindenau in Von Zach, Monat. Corresp., 1806, Vols. XIII-XV, but which has been studied by proper analytical methods only within the past ten years by Oberbeck and Helmholtz. I shall attempt to obtain general results only as to the influence of the sun upon the growth of clouds and storms. We may analyze the insolation into that which is directly absorbed by the atmosphere and that which finally reaches the ground and ocean. Of that which is absorbed in the atmosphere (30 or 40 per cent.?), one-half is radiated back into space, the other half towards the earth.

Now radiation from a warm to a cold substance increases with the difference of temperature, and in Fig. 59 I have shown the general character of the rate of radiation for several cases in which we are interested: Curve 1A shows the diurnal curve of rate of cooling of the earth's surface into space for two cases, first, that in which the sky is absolutely transparent to solar and terrestrial radiations; curve 1B, that in which the sky absorbs four-tenths of the zenithal solar ray, but none of the terrestrial radiation at night time, which latter feature of the absorbing power of the air accords nearly with Langley's newest results; curve 2 shows the rate of warming of the soil by its absorption of solar radiation for an average transparency co-efficient of 0.6; curve 3 shows the diurnal curve of rate of warming of the lower stratum by its absorption of the radiation to it from the upper atmospheric layers;

curve 4, total rate of warming of the lowest atmospheric layer by radiation; curve 5, the rate of change of temperature of the lowest surface of the air due to contact with the surface of the ground (and, therefore, having the same temperature as it), for two co-efficients of absorption of the solar rays, namely, 1.0 for perfect transparency and 0.6 for ordinary partly cloudy weather; curve 6*A* and 6*B*, the resulting hourly rates of change of temperature for the two cases *A* and *B*; finally, the sums of these rates of changes and the integrated resulting curves of temperature of the lower air. The general effect of the increasing transparency is seen to be the rapid increase in the extremes of the diurnal fluctuations of temperature, and on these extremes depend the diurnal variations in vertical convection currents, in variations of the wind, and in its frequency and violence of local storms.

In Figs. 60, 61, 62, I have given for latitudes 30, 40, and 50 degrees the quantities of heat received during any day as computed by Angot (See Annales de Bureau Centrale de Météorologie de France, 1883, Tome I, Plates B10 and B11) for different values of the co-efficient of transparency, namely, for $p = 1.0$ and 0.8 and 0.6, the latter co-efficient corresponds to the loss experienced by the sun's rays in average weather in northern latitudes, the second co-efficient corresponds to our very clearest weather, but the first co-efficient or unity corresponds to the case of no loss by absorption, that is to say, to the actual total heat received at the outer surface of the atmosphere.

The difference between the curves for 1.0 and 0.6 represents the heat lost in, namely, absorbed by the atmosphere, and this immense quantity, much of which falls upon the tops of the clouds, exerts a large influence in the atmospheric motions whose importance meteorology has as yet scarcely begun to consider. The curves representing this difference are given on Fig. 63. In all these figures a scale of relative units is given on the right-hand side and a scale of absolute calories on the left-hand side, in which latter I have assumed the solar constant to be 3.0 calories per square centimeter per minute.

I have already, in the article on the diurnal variation of the barometer, briefly mentioned that Dr. Hann proposes to explain that phenomenon as the direct result of the absorption of solar rays by the upper strata, but in my own opinion the heat thus retained above in the upper layers must affect the barometer not directly, but indirectly; its direct effect is to alter the buoyancy, but its immediate indirect effect is to alter the motions of the atmosphere, and through these to alter the barometric pressure as observed by us.

8. The total amount of heat received during any one day has been computed by many, beginning with Lambert; the most recent memoirs are those by Angot and by Zenker, between whom the prize of the Paris Academy of Sciences has been equally divided during the past year; as yet the first of these memoirs is known to me only through

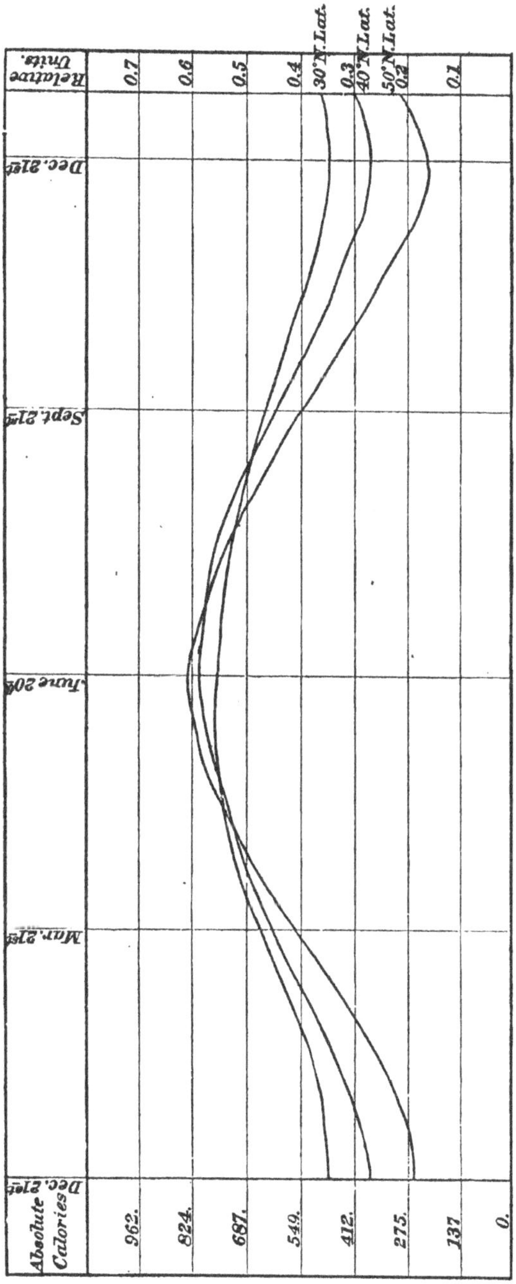

FIG. 62.—Heat directly absorbed by the atmosphere and clouds when $p = 0.6$.

notices and reviews, and the previous memoir by Angot, published in the Annales of the Central Meteorological Bureau of France for 1883, is the one whose results are given in a graphic shape, most convenient for my use, therefore, the preceding diagrams (see Figs. 60 to 63) have been based upon the work of Angot. From these we can read off, by means of the scale of absolute calories, the amount of heat received at any day of the year at any latitude, and in ordinary conditions of the sky, either at the upper surface of the atmosphere or at the ground; thus, from Fig. 60, we see that on the 20th of June the upper surface of the atmosphere receives in twenty-four hours 1,512 calories, while the ground receives during ordinary fair weather 690 calories; therefore, during the whole day, the difference between these numbers, or 822 calories, has been directly absorbed by the air and the clouds, or very much more than actually reaches the ground. But this total absorbtion during the day has been very unequally distributed through the hours. The ratio of the absorbed to the transmitted heat is greatest when the sun is lowest, and least when the sum is at the zenith, and this fact causes the figures, which we have above given for June 20, to present a very different ratio among themselves from those that would be given by the same diagram for the date December 21; thus, on this latter day, for latitude 30° N., Angot's diagram (see Fig. 60) gives for the outer layer of atmosphere 700 calories, and for the earth's surface 220 calories, the difference between which, or 480, shows that the quantity of heat directly absorbed in the atmosphere is on this date more than twice that which reaches the ground [instead of $\frac{822}{690} = 1.2$ as on June 20], and this ratio increases as we go northward, although the absolute quantities of heat rapidly diminish, so that on December 21 at 50° N. latitude we have from Fig. 62, for the ground 20 calories, but for the upper surface 260 calories; the difference, or 240, shows that the direct absorption is eleven times the quantity of transmitted heat. This ratio, depending as it does on the annual change in the declination of the sun, has its parallel in the hourly proportions of absorbed and transmitted heat depending on the diurnal change in declination of the sun. In order to shorten the labor of using such diurnal curves I, in 1871, prepared charts, which were used for several years and finally submitted to the Chief Signal Officer, showing approximately and for the 1st and 15th of each month the relative amount of heat that is received at any point in the United States during the interval between the successive tri-daily charts of the Signal Service on the assumption of an average co-efficient of absorption of 0.7. The general result shown by such charts may also be deduced from a study of the tables given by Meech, Angot, Ferrel, and other writers on this subject.

It is, however, desirable to have some method, even if it be a crude one, of determining by observation the amount of heat transmitted or absorbed under the prevailing condition of the sky as to haze, cloud, etc. This has been attempted to be attained by the records of the white and black bulb in vacuo. For a perfectly clear sky of uniform and homogeneous constitution the difference, black bulb minus bright bulb, should be very nearly the same for a given altitude of the sun, morning or evening, but, like all other actinic measures, this apparatus shows that there is a greater absorption in the afternoon than in the morning, a greater in moist warm weather than in cool dry weather, greater in summer than in winter, greater in equatorial regions than in polar regions, and greater in the southern hemisphere than in the northern hemisphere. These results all point to the conclusion that an increase in the moisture contained in the atmosphere increases the absorption, but they do not of themselves enable us to distinguish between the effect of the moisture that is present as invisible vapor and that which is present as haze or cloud; on the other hand, the whitish appearance of the sky which we usually call light haze and which corresponds to the lower numbers of the 51 divisions on the scale of Saussure's Cyanometer shows that in moist weather or in equatorial climates the moisture effective in the absorption or dispersion of heat is that which is present as most finely divided water or the aqueous globules that are really ultra-microscopic but whose diameter can be approximately determined by optical methods. It is this comminuted water that absorbs the rapid vibrations of the blue end of the spectrum, degrades the ultra blue vibrations down to the blue, and diffusely reflects the red or ultra red vibrations so that the observer under a hazy sky receives from the whole hemisphere above him the very appreciable amounts of heat shown in the fifth column of Zenker's table. (See Zenker, Die Vertheilung der Wärme auf der Erd-Oberfläche.)

Now these long waves thus received by the black bulb and bright bulb *in vacuo* do not pass through the glass inclosure that surrounds these bulbs, and the readings of the Arago-Davy actinometer are able to give us only an idea of the heat effect due to those radiations that can penetrate the glass, consequently the value of the so-called constant of solar radiation will vary with the quantity and nature of the haze, and, therefore, in general with the time of day, the season of the year, the latitude, and the continental and oceanic position; therefore, the indications of this instrument can only serve as a rough method of deducing the current distribution of solar heat; its error, in this respect alone, to say nothing of other sources of error, may be estimated at 20 per cent. in the northern latitudes and fully 30 per cent. in equatorial oceanic regions, where the degraded and diffused radiations of long wave length are, on account of the hazy sky, much more important than in northern regions. The estimate above given is my own conclusion from a study of the valuable observations made by Professor Upton in May, 1883, at the Caroline Island, and recorded in Volume II of the Memoirs of the National Academy of Science. These same observations have also been discussed by Professor Ferrel in the Signal Service Professional Papers No. 13, and in his Recent Advances, pages 131 and 378, where he has shown the effect of sluggishness in the thermometers as affecting the resulting computed intensity. Professor Ferrel also gives his results from his own observations made at Washington in March and May, from which we see the effect of irregularity produced by the slightest haze. From all this I conclude that if the solar intensity of solar radiation as known approximately from other sources could be adopted as a normal value, and if then the intensity be observed by means of the bright and black bulb at any moment (which gives us only the effect of the higher wave lengths) then the difference between this latter quantity and the former gives a measure of the amount of heat absorbed and diffused and degraded as to wave lengths by the atmosphere at that moment.

9. Another method of approximating to the quantity of heat respectively transmitted and absorbed is based on the consideration that the cumulus clouds as well as all the thicker clouds from which rain falls absorb a very large per cent. of all the heat that falls upon them, and the same is true of layers of fog or mist; the heat that actually penetrates such cloud or fog is considered as made of two parts, a few of the longest wave lengths, such as can possibly directly penetrate water or for which the co-efficient of transparency is near unity; second the much larger quantity of heat that by diffused reflections and refractions at the surfaces of the globules of water is able to penetrate the cloud and reach the earth below. These latter reaching the earth at any spot from all directions namely from the whole mass of cloud or fog are not parallel rays, and can therefore not be concentrated by any ordinary lens upon a definite spot; these are therefore not taken account of in photographic records of the intensity of daylight or in the records made by Campbell's Sunshine Recorder.

Let us consider therefore the solar rays as effective when they penetrate a cloud directly or when they penetrate the blue sky between the clouds; these rays being parallel to each other therefore the alternate cloudy and clear spots in the heavens are protected by parallel rays upon the earth's surface, so that the proportion of shaded and illumined regions on the earth's surface are for any locality the same as the observed proportion of clear and cloudy sky vertically above us. Let these proportions as measured in horizontal area be a and b, then the whole area $a+b$ and the illumined area b are to each other in the ratio $\frac{b}{a+b}=\frac{1}{1+\frac{b}{a}}$. This ratio is, however, for any given altitude of the sun the same as the ratio of time-duration given by the ordinary sunshine recorder by photography; moreover if Campbell's Sunshine Recorder be used, or Marchand's apparatus, which he calls the photantitupimeter, then by the depth of the charred wood, or by the chemical work done, we get a fair approximation to the intensity of the solar rays.

If one has a photographic record only of the duration of sunshine, then the total amount of heat received at the earth's surface must, if we would be accurate, be computed for each hour of the day and summed up for the whole day, on the assumption that the heat received during the actual hour is to the total possible heat during a clear hour in the ratio of the duration as shown by the photograph to the maximum possible duration.

In the absence of photographic records I have been accustomed to make use of the tri-daily observations as telegraphed, showing the kind of clouds prevailing at each station and the quantity of cloudiness. In a general way the diurnal periodicity of cloudiness is sufficiently well marked to justify us in the saying that if the 7 a. m. chart shows zero or 10 per cent. of clouds, and the 3 p. m. 80 to 100 per cent. of cumuli, then the 11 p. m. chart would have shown these latter to have almost entirely disappeared. Assuming that the 3 p. m. record gives us the maximum cloudiness for the day, and that these clouds are thick enough to absorb all the heat that falls upon them, we are at once able to say that of the solar heat not absorbed in the clear air in general the percentage that falls upon and is absorbed by the clouds is the same as the reported percentage of cloudiness.

10. The buoyancy of the air varies with its transfer northward or southward, since such motion brings it into other latitudes where the heat that it will receive from the sun differs from that to which it has been accustomed. The air that in a southern latitude was maintained clear of haze by the solar rays will cool and form haze when transferred northward, for the reason that, while its radiating power remains the same and its moisture remains the same and its pressure has changed but little, yet the heat received from the sun and needed to keep that moisture in the state of invisible vapor has diminished. Inversely in air that moves southward, as in our great waves of cold air and high pressure, the moisture remains nearly the same, the pressure generally diminishes, the slight consequent expansion produces a slight diminution of temperature; but this is more than counteracted by the increasing insolation, as well as the radiation from warm earth and the evaporation from water and snow surfaces, so that the temperature rises, the relative humidity becomes less, and the formation of cloud or haze diminishes. The following table (I) shows the loss or gain of heat in calories per day so far as the sun's direct rays are concerned as received by a square centimeter of horizontal surface at the upper layer of the atmosphere and again at sea-level under a fair sky when transported 10 degrees northward from any given latitude on the 15th of any month. The figures are very nearly the same, with the reversed sign for the same surface when transported southward:

TABLE I.—*Effect of moving a unit of horizontal surface 10° northward.*

CO-EFFICIENT OF TRANSPARENCY 1.0.

Month.	0-10	10-20	20-30	30-40	40-50	50-60	60-70	70-80	80-90
January	−133	−155	−179	−190	−191	−179	−113	− 4	0
February	− 86	−121	−143	−171	−183	−191	−177	−107	− 4
March	− 19	− 59	− 93	−120	−153	−170	−187	−193	−121
April	+ 37	+ 8	− 28	− 61	− 90	−121	−128	− 94	+ 1
May	+ 90	+ 58	+ 27	− 4	− 35	− 59	− 31	+ 43	+ 19
June	+109	+ 82	+ 50	+ 20	− 5	− 18	+ 38	+ 63	+ 23
July	+ 93	+ 70	+ 31	+ 8	− 14	− 32	− 8	+ 58	+ 20
August	+ 55	+ 25	− 14	− 47	− 77	− 97	−190	− 47	+ 8
September	− 8	− 39	− 74	−105	−136	−159	−175	−187	−109
October	− 70	−101	−133	−155	−175	−191	−187	−140	− 23
November	−120	−152	−167	−187	−191	−182	−132	− 16	0
December	−144	−167	−187	−194	−191	−167	− 74	0	0

CO-EFFICIENT OF TRANSPARENCY 0.6.

Month.	0-10	10-20	20-30	30-40	40-50	50-60	60-70	70-80	80-90
January	−90	−105	−111	−110	− 85	− 39	− 4	0	0
February	−58	− 85	−101	−113	−101	− 82	− 35	− 4	0
March	−19	− 47	− 70	− 93	−105	−110	− 93	− 50	− 8
April	+17	− 4	− 31	− 58	− 82	−101	−113	−101	−50
May	+59	+ 31	+ 7	− 27	− 50	− 79	− 89	− 85	−32
June	+74	+ 46	+ 20	− 8	− 35	− 59	− 78	− 57	−17
July	+66	+ 39	+ 11	− 19	− 47	− 70	− 85	− 75	−23
August	+36	+ 8	− 23	− 47	− 74	− 94	−109	−100	−55
September	− 3	− 35	− 63	− 85	−101	−105	−102	− 70	−15
October	−50	− 78	− 94	−105	−109	− 89	− 50	− 8	0
November	−84	−101	−113	−105	− 90	− 50	− 8	0	0
December	−97	−113	−113	−101	− 78	− 27	0	0	0

Table II* shows what the unit surface at sea-level exposed normal to the sun's rays all day long (*e. g.*, a globe or cloud in the full sunlight) would lose by being transported 10° north. In this case the diminution of heat received is due to atmospheric absorption only. In the former case it is due not only to this absorption, but to the increased inclination of the horizontal surface to the sun's rays.

A northward movement therefore conspires with the diurnal radiation to accelerate the production of haze and clouds, but a southward flow retards the production of haze. The actual statistics as to the exact amount of this effect in comparison with the other causes that are at work to produce cloud have yet to be collected, but in a general way the study of successive weather maps shows that of two streams of the same air flowing from a given central region simultaneously northward and southward, the one flowing northward will be filled with haze or cloud about twice as soon as regards time and about three times as soon as regards distance as the stream which is flowing southward.

* Omitted.

11. The formation of the cloud depends entirely on the fact that aqueous vapor is present in the air, by reason of whose lightness and latent heat the active formation of ascending currents and clouds is determined. The process by which the lower stratum of air becomes more or less perfectly saturated with moisture, namely, the process of evaporation, is one that requires careful consideration. The evaporation from moist surfaces depends upon the quantity of moisture available at the surface, and for almost all surfaces of plants and earth it is less than from the surface of fresh water. So, also, is the evaporation from salt water only one-third or one-half that from fresh water; but the latter is less than from the surface of freshly fallen snow. The evaporation from fresh water is that which has been most generally studied and found to depend essentially on the velocity of the wind, the pressure of the atmosphere, the temperature of the surface of the water, and the temperature and dryness of the air blowing over that surface. The evaporation consists of a comparatively slow diffusion of vapor particles into the adjacent air, which diffusion would soon saturate that air but for the convection due to horizontal wind and to rising convection currents from the surface of water.

The relation between the actual evaporation and the conditions upon which it depends have been expressed by several students in formulæ whose constant co-efficients are deduced from observations on the evaporation of pure water in shallow dishes set up within a shelter similar to a thermometer shelter, and, therefore, protected completely from solar radiation and partly from the full force of the wind; under these conditions the evaporation must be much less than would take place from the same dishes exposed in open fields to the full sunlight and the wind; but this latter form of exposure would certainly give a larger evaporation than actually takes place in nature. As the evaporation in the open air from the dry ground and the leaves of plants is less than that from freely exposed pure water, therefore the diminished evaporation observed in the sheltered dishes has been accepted as a crude approximation to the evaporation that actually takes place in nature, but strictly speaking it represents only the maximum power of the shaded air to evaporate, and in a dry continental locality this evaporating power of the air is far in excess of the power of the soil and plant to supply the needed moisture; some plants evaporate much more freely from their leaves than do others and the relative powers have been experimentally ascertained, but for our purposes we are obliged to consider only the evaporation from the free surface of water or by preference from the free surface of moistened cloth or paper as used in the ordinary psychrometer or in the Piche evaporometer.

The relations established by different experiments may be briefly summarized as follows:

(1) Mr. Thomas Tate, L. E. and D., Philosophical Magazine, 1862 (4),

Vol. XXIII, page 126, Vol. XXV, page 331, as the result of some experimental researches deduces the following laws:

(*a*) Other things being the same, the rate of evaporation is nearly proportional to the difference of the temperatures indicated by the wet-bulb and dry-bulb thermometers;

(*b*) Other things being the same, the augmentation of evaporation due to air in motion is nearly proportional to the velocity of the wind;

(*c*) Other things being the same, the evaporation is nearly inversely proportional to the pressure of the atmosphere;

(*d*) The rate of evaporation of moisture from damp, porous substances of the same material is proportional to the extent of the surface presented to the air without regard to the relative thickness of the substances;

(*e*) The rate of evaporation from different substances mainly depends upon the roughness of, or inequalities on, their surfaces, the evaporation going on most rapidly from the roughest or most uneven surfaces; in fact, the best radiators are the best vaporizers of moisture;

(*f*) The evaporation from equal surfaces composed of the same material is the same, or very nearly the same, in a quiescent atmosphere, whatever may be the inclination of the surfaces; thus a horizontal plate with its damp face upwards evaporates as much as one with its damp face downwards;

(*g*) The rate of evaporation from a damp surface (namely a horizontal surface facing upwards) is very much affected by the elevation at which the surface is placed above the ground;

(*h*) The rate of evaporation is affected by the radiation of surrounding bodies;

(*i*) The diffusion of vapor from a damp surface through a variable column of air varies (approximately) in the inverse ratio of the depth of the column, the temperature being constant;

(*j*) The amount of vapor diffused varies directly as the tension of the vapor at a given temperature, and inversely as the depth of the column of air through which the vapor has to pass;

(*k*) The time in which a given volume of dry air becomes saturated with vapor or saturated within a given percentage is nearly independent of the temperature if the source of vapor is constant;

(*l*) The times in which different volumes of dry air become saturated with watery vapor (or saturated within a given per cent.) are nearly proportional to the volumes;

(*m*) The vapor already formed diffuses itself in the atmosphere much more rapidly than it is formed from the surface of the water (this assumes of course that there are no convection currents of air to affect the evaporation or the diffusion).

In order to express the evaporation from water surfaces in numerical formulæ Weilenmann has elaborated the original formulæ of Dalton (see Swiss Meteorological Observation, 1875, Vol. XII), and finds that the monthly evaporation as measured by the depth of water evaporated from sheltered dishes of pure water can be expressed by a formula which after dividing by thirty gives the evaporation per diem, as follows:

$$v=\mu_1\left(\Sigma\frac{m}{\alpha+\lambda}+\gamma\Sigma\frac{mV}{\alpha+\lambda}\right)$$

where v is the daily evaporation, m is the deficiency of vapor present in the free air before evaporation, namely, the difference between the weight of vapor expressed in grams per cubic meter for saturated air at the current temperature less the weight actually present; this difference in weight $m=g'-g$ is also the same as the difference in vapor tension $e'-e$ multiplied by a constant factor.

V is the velocity of the wind at the surface of the water in kilometers per hour, μ_1, λ, and γ are constants to be determined from observations.

On applying his formula to all available observations Weilenmann obtains the following results for diurnal evaporation:

For Vienna $v=0.673$ millimeters $\times(s'-s)$, where s' and s are diurnal averages, and the average wind effect is included in this one term.

For St. Petersburg the same formula holds good with the co-efficient 0.675.

For Mount Souris, where the wet paper disc of the Piche Apparatus was the evaporating surface, the co-efficient was 0.679 and the second term depending on the velocity of the wind was appreciable but small.

For Pola the co-efficient was 0.726 and the second term was inappreciable.

For Tiflis the co-efficient in this very dry climate varied exceedingly, namely, between the limits 0.3 in winter and 1.4 in summer, the average being 0.7.

For Moncalieri the average value of the co-efficient is 0.67 with a small appreciable term depending on the wind.

From the preceding we conclude that the evaporation from natural surfaces sheltered from sun and wind is accelerated by light local currents to such an extent that the slight additional acceleration produced by the effect of ordinary wind is scarcely appreciable.

In order to ascertain something with regard to the effect of direct sunlight and the free winds I quote the results of observations at Tiflis, by Noeschel, as given in Vol. V of Wild's Repertorium. His result is that during the storm months the evaporation during the twelve hours from 8 a. m. to 8 p. m. is, in the open air, three times that in the shelter; from 8 p. m. to 8 a. m. the ratio is 1.8, and for the whole twenty-four hours the ratio is 2.6.

Stelling, from a discussion of all observations accessible to him (see Wild's Repertorium, Vols. VII and VIII), finds that Weilenmann's formula is generally applicable, but of course the constant co-efficients must vary, especially with the latitudes and the point at which the anemometer is located, for the purpose of determining the relative velocity of the wind; thus, for observations near St. Petersburg, at Pavlosk, he finds

$$h=0.274\Sigma(S-s)+0.0317\Sigma(S-s)w$$

where w is the velocity of the wind in meters per second as measured by an anemometer 28 meters above the evaporating surface; S and s are the vapor tensions for the temperature of the air and the dew-point; h the measured depth of the evaporation in millimeters per day. The evaporometer was a dish floating in a tank of water in the open air.

For a similar dish set in the ground at St. Petersburg Stelling finds the co-efficients to be respectively 0.3923 and 0.045.

For a free dish of water in the sun, but not floating in a tank, in the dry climate of Nukuss and for the summer season, Stelling finds the co-efficients 0.8124 and 0.1378. These latter are, therefore, three or four times as large as the corresponding co-efficients at Pavlosk.

From observations at Nukuss on dishes sheltered from the sun and the wind the co-efficients deduced by Stelling, when converted into the units of measure above given, become, respectively, 0.552 and 0.170.

In the United States observations have been made by Fitzgerald at the Chestnut Hill Reservoir, near Boston, and he states that his results can be represented by the formula

$$E=0.0166\,(V-v)\,(1+\tfrac{1}{2}w)$$

where E is the evaporation in inches per hour; V and v are the tensions of vapor in inches for the temperature of the air and the dew-point, respectively; w is the velocity of the wind in miles per hour at the surface of the water as obtained by assuming that the anemometer used by him and located at a little distance gave three times the velocity desired.

The co-efficient given by Fitzgerald represents the average of the year, and the variations of vapor tension and wind give the annual and diurnal variations in the quantity of evaporation.

The direct effect of wind on evaporation was, in spite of the preceding investigation, only very imperfectly known as a residual effect after temperature and dryness had been allowed for until the accurate observations made by Prof. Thomas Russell, of the Signal Service, in 1887, gave us a few precise measurements which, however, still need to be repeated at many other temperatures.

The following data are taken from the Signal Service Monthly Weather Review for September, 1888, page 176. The Piche evaporometer was whirled at different velocities on a whirling-machine through the still air of a large room, and a similar instrument was simultaneously observed at rest, hence resulted the following relative weights or depths of evaporation from the wet surfaces of the paper disks. The temperature of the air in the room was 83.7°F., the dew-point was 63.5°, the relative humidity 50 per cent., the vapor tension for the two temperatures is 1.15 inches and 0.58 inches, the reading of the wet bulb of the whirled psychrometer was 70.5°F.

Professor Russell's observations gave the data in the first two columns, to which I have added a third, showing the effect of a mile of wind in affecting the evaporation at any given velocity.

Relative evaporation from Piche for different winds:

Velocity of wind per hour.	Relative evaporation.	Effect of change of 1 mile in the velocity.
Miles.		
0	1.0	....
5	2.2	0.24
10	3.8	0.32
15	4.9	0.22
20	5.7	0.16
25	6.1	0.08
30	6.3	0.04

Further computations are given by Professor Russell with charts, showing the distribution of evaporation over the United States, from which, however, only very general conclusions can be drawn as to the actual evaporation taking place during a given day or a given condition of the weather.

It may perhaps in general be assumed that one-fourth of the water that falls as rain and one-half of the water that falls as snow is evaporated from the ground and again precipitated as rain or snow, until, after several repetitions, it reaches the ocean and ceases to be recorded as rain-falls on the continents. In this way each locality on the land is made to supply moisture to the air flowing over it. The process by which this moisture rises up to form clouds and rain is precisely similar to that by which heat is conveyed by rising currents of air from the ground to the upper portion of the atmosphere. The path of a stream of particles of vapor is like that of the current of warm air approximately depicted in Fig. 64, where we see moist air rising at *A* and floating with the wind eastward toward *B*, over which it has risen high enough to form a cloud from which rain is falling; the sun shining on the western side of the cloud gives it a buoyancy and an especial ascending motion, while the eastern side of the cloud is in the shade; the rain descends from the colder region, and which is also not possessed of too violent upward currents, and therefore especially from the regions *C D* on the shady side of the central portion. The ascent of the stream of vapor over the region *A* is at first nearly vertical because of the slight winds near the ground, but after attaining an altitude of 100 feet the horizontal movements become more rapid and the vertical movements slower; moreover in general, as before stated, the ascending masses acquire a rotary motion, and are liable to break up into numerous smaller rotating vortices. The horizontal distance covered by our diagram must be considered to vary exceedingly with the season and locality and the hour of the day; thus, in many summer thunder-storms masses of rising air have been observed to form cloud and rain at a height of 3,000 to 5,000 feet within a distance of from 20 to 50 miles, and within a time estimated at from less than one up to three hours; in other cases undoubtedly the ascent takes place much more slowly, as, for instance, in the case of evaporation during strong northwest winds or cold waves, when the evaporated moisture may be carried 1,000 or 2,000 miles and not begin to assist in the formation of clouds until several days after starting.

12. When by the ascent of moist air the moisture is eventually condensed to form haze and cloud, there begin two distinct thermal processes:

(*a*) The evolution of latent heat of liquefaction by which the cloudy-rising air is made warmer and lighter, and therefore more buoyant than it would have been, and therefore rises all the more rapidly;

(*b*) The general absorption by the cloud of all solar radiation that falls upon its upper surface;

(*c*) The absorption by the cloud of all the terrestrial radiation that strikes its lower surface;

(*d*) The absorption of all the atmospheric radiation that strikes the cloud from all sides. In these two last categories we have to omit from consideration the almost infinitesimal special radiations to which the cloud may be transparent;

(*e*) The change in radiation from the cloud particles as compared with the radiation from the invisible vapor within the cloud.

The preceding items, in so far as they deal with the radiation and absorption from gases, have to do with a subject that has not yet been very satisfactorily handled by physicists, but so far as meteorological phenomena are concerned we may, I think, safely proceed on the following approximate laws:

A gas or vapor is opaque to or has a small co-efficient of transparency for the radiations emitted by itself when in precisely the same condition as to temperature, pressure, and moisture, but when the conditions of the two masses are very different in these respects the absorbing power of the denser, warmer, or more humid mass is the greater. It thus happens that although the lower atmosphere can radiate a little outward through the upper, yet the upper air finds its radiations downward almost wholly absorbed by the lower atmosphere. We shall assume that a cloud absorbs all the radiation that falls upon it, and that it returns in longer wave-lengths or degraded radiation a small part only of its heat from the upper surface or summit, and loses no appreciable amount from the sides or bottom over and above what it receives on these sides.

In the rolling and overturning of the air as it ascends to form a cloud one portion must be descending while the succeeding portions are ascending, so that eventually all the air that has ascended in the cloud descends to lower levels and back to the earth's surface. In the case of rapid motions, as in tornadoes and thunder-storms, an appreciable amount of ascending air is by its inertia carried up above the position of static equilibrium, and that portion of it that is not brought down by rotary motions must at this high level mix with surrounding air and float away horizontally, descending a little so as to form the horizontal broad clouds that stream from the tops of cumuli, and that are most beautifully illustrated in the clouds that surmount the volcano Vesuvius when in active eruption.

The radiation from the upper surface of the cloud is of very much the same character as that from moisture at the surface of the earth, as experimented upon in our laboratories, and consists principally of long waves. Owing to the finely divided character of the moisture the coefficient of emissivity for these waves is nearly unity. The heat lost at the upper surface is moreover a maximum because the minute radiating surfaces as soon as they cool must fall slowly and be replaced by other rising slightly warmer particles. The mixture continually taking place at the surface of the cloud, with the cooler, drier air surrounding it, as explained in the chapter on vortex motions, together with the radiation into space and the radiation from the sun, combine to determine the shape of the upper surface. In general, the heat from the sun causes the illumined half of the cloud to be lighter, and therefore to rise up steeper, as shown in Fig. 64, where the sun shining on the west side has developed that half of the cloud into a special system of vertical currents, while the opposite shaded side of the cloud, having much less buoyancy, extends more to the eastward and remains at a lower altitude; therefore the sunny side is steep and the shaded side slopes gradually. When the cloud, however, is below other clouds and wholly or partly shaded, then the temperature of the upper surface is not lowered by radiation nor heated by the sun, and the warmth of its own vapor serves to maintain a very slow ascent and the cloud assumes the shape of a thin horizontal sheet.

The temperature of the upper surface of the cloud as compared with the temperature at the earth's surface differs from the latter in this respect, that the surface layer of the cloud is movable, while that of the earth is fixed. New particles renew the cloud surface and keep its temperature more nearly uniform. The heat supplied to the surface of the earth during the night-time is principally by conduction from below, while that supplied to the upper surface of the cloud is convected from below; therefore the temperature of the earth diminishes steadily during the night, while the temperature of the upper surface of the cloud will not diminish below the dew-point of the cloud at that altitude, except in so far as the dew-point of the whole cloud is uniformly reduced by precipitation. Therefore the temperature of the upper surface is very approximately that of the dew-point as given by Hertz's graphic table.

As the relative buoyancy of the upper surface of the cloud with regard to the adjacent atmosphere on its own level depends on the temperature of the latter, this has therefore been the subject of considerable discussion. We may attain some idea of the temperature of the free air at great altitudes from balloon voyages, such as those of Glaisher, and from phenomena of atmospheric refraction; all other methods are based upon theoretical extrapolations, which may start with observations at the lower levels or may depend upon some theory as to the temperature at higher levels. If we pass from lower levels upward we have virtually to assume that the air at high altitudes has at some past time within the previous two weeks risen to those altitudes by a very gentle gradient, starting from some point many miles away, and is therefore a different and a drier mass of air, and has a slightly lower temperature, or at least is of slightly greater density, than the air within the rising cloud.

The general law of diminution of temperature for such masses of air is fairly given by the equations established by Mendelief, Hann, G. and F. Chambers, and others. If, on the other hand, we can make any reasonable assumption as to the temperature at the outer portion of our atmosphere, we may from that point reason as to the temperatures of strata that are nearer the ground and at the upper level of the clouds; it has been customary to assume that the outer layers of the atmosphere are at the temperature of space, therefore very near the absolute zero; but this hypothesis needs revision in view of the consideration that at all altitudes above 15 or 20 miles the air must be transparent for long waves, and can only absorb short waves, if any; that therefore it must receive all of the heat that it retains by the direct absorption of the solar rays, and however slight this absorption may be it must elevate the temperature decidedly above that of any ulterior space, if such there be, where none of the ordinary forms of matter are supposed to exist. But in these higher portions of the atmosphere there must be retained an appreciable percentage of the lighter gases that have from time to time escaped from the earth's surface, and, as it can scarcely be thought that such lighter gases participate in the convectional interchange with the heavier gases below, therefore their temperature is a matter that can not be inferred from observations made in the nitrogen, oxygen, and aqueous vapor of the lower atmosphere. The temperature of this lighter layer is that uniform temperature that controls our own convectional atmosphere, and I see no *a priori* reason why there may not exist at these high elevations regions of warmth which shall lose their heat only very slowly by radiation, or regions of great cold which shall very slowly warm up by the absorption of solar radiation. These latter regions descending and mixing with the air at elevations of from five to fifteen miles would contribute powerfully to the maintenance of clear sky.

The transfer northward of warm moist tropical air, which of course ascends slowly and overlies the colder air of the temperate regions, should give rise to a phenomenon that we have come to designate as warm waves. The mechanism of these is somewhat as follows: The warm air borne northeastward on the upper southwest current would always stay above the colder air near the ground, except during the latter portion of the day, in which the surface air is highly heated by the sun; on such occasions and in such localities the ascent of the hot surface air brings down from above an air that, although denser than the ascending, is still quite warm; the mixture of the two gives us an air that is cooler than that prevailing at the very surface of the earth but relatively much hotter than would have been the case had the upper air thus brought down been of the lower temperature that ordinarily prevails 1 or 2 miles above the earth's surface. The process of mixture that takes place in the interchanging convection during a warm wave is very different from the process of compression and evolution of latent heat that takes place without any great mixture, as in the phenomenon of the foehn wind of Switzerland or the hot winds of Montana and the Argentine Confederacy. If, in these latter winds a large percentage of intermixture could take place, the heat of the descending currents would be tempered and they would become warm waves instead of hot winds.

During the winter time both Europe and America and Asia are subject to so-called cold waves; it is not likely that these can be due merely to the descent of cold air from above; all the circumstances point to the conclusion that they represent the horizontal flow of immense masses of dry air from northern regions, and that their cooling has been originally due to terrestrial radiation unopposed by solar heat, aqueous vapor, or cloudy skies. They therefore give us no clue as to the temperature of the upper atmosphere; in fact, they are themselves comparatively shallow and broad masses of cold dry air, whose dryness is as important as their temperature. The advent of the southern limit of such a cold wave is usually preceded in the United States by the appearance of a low barometer to the west of Oregon and Alaska, the flow of air southward is greatly favored by the presence of the immense flat area of Hudson's Bay and the surrounding low lands; the flow of air westward is hindered by the Rocky Mountain Range, and its flow southward is largely determined by the formation of cloud and storm on its southeastern border.

To return now to the buoyancy of the cloud whether formed by a special local ascending current, as in individual cumuli, or by the general ascent of warm air along the front edge of a cold wave, we see that in either case the heat lost at the upper surface of the cloud by its own radiation must be that which determines the rain-fall, or if any rain falls by reason of rapid condensation within the cloud before it has time to be lost by radiation that heat eventually permeating the cloud must be subsequently lost by radiation, so that in general the loss of heat by radiation and by rain-fall are mechanically equivalent. The falling of the rain is a rapid process, but the radiation of the heat comparatively slow. In an interim while the heat is being convected through the cloud and lost by radiation the cloud derives the advantage of the buoyancy due to its presence, and thus a storm is maintained while passing over dry regions or while the sun is absent at night-time.

13. In order to completely determine the buoyancy of a given layer of cloud, we have to consider not only the temperatures at the lower surface of the cloud, as just determined, but also, as far as practicable, three other items, namely: First, the thickness of the cloud; second, the effect of insolation upon the upper surface; third, the effect of heat left in the cloud by vapor that has fallen as rain.

(*a*) The first of these is approximately given by a consideration of the darkness of the cloud-covered sky, and in some cases even a photometric determination of the amount of sunlight that penetrates the layers of clouds has been made. See the work done by Leonhard L. Weber at Munich on the occasion of the solar eclipse of August, 1887. The methods used by him would be applicable to general meteorological observations if at any time it seemed practicable to introduce the estimate of cloud thickness and buoyancy into the determination of movement of storms. For the present an estimate by the observer on an arbitrary scale of tenths of the thickness of the cloud layer is all that is practicable.

(*b*) The effect of insolation upon the upper surface is found by first summing up the amount of heat received at the upper surface between the dates of the successive weather charts, according to the methods already provided, assuming that the atmospheric absorption by the atmosphere above those upper layers is negligible, and that the heat reflected and radiated from the cloud surface is about the same as that reflected and radiated from the ocean. In this case the relative heat received during any minute by a unit of cloud surface, increasing, as it does, with the altitude of the sun, is very nearly represented by the following table:

Relative heat received by unit of cloud surface.

Sun's Z. D.	Incident heat	Absorbed heat
0	0.96	0.93
10	0.95	0.92
20	0.90	0.87
30	0.83	0.80
40	0.73	0.70
50	0.60	0.57
60	0.46	0.43
70	0.30	0.27
80	0.14	0.12
90	0.02	0.01

If the unit of heat is the 3 calories of solar radiation received at the outer surface of the atmosphere, and if 0.86 of this penetrates the zenith to the cloud at an assumed altitude where the barometric pressure is 20 inches, and if the sky reflection adds 10 per cent., then a zenithal sun gives the top of the cloud 0.96 of 3, or 2.88 calories per minute per square centimeter.

The relative percentages of total solar heat received at the outer surface of the atmosphere in the intervals between the successive tri-daily maps vary with longitudes and latitude and are given in the following table, the unit being the total diurnal amount for the given latitude and date at the outer surface. For greater definiteness I have added in the last two columns the values of this diurnal unit for each latitude expressed in gram-calories per horizontal square centimeter. This value in absolute heat units is subject to all the uncertainty attending our present knowledge of the constant c of solar radiation, for which we have three prominent authorities, viz.: Pouillet (c=2.18), Violle (c=2.54), Langley (c=3.00); whence result three values of the heat received by unit horizontal surface in one day at the equator, viz., 966, 1164, or 1375 calories:

JUNE 20.

West longitude.

Latitude (north).	11 p. m. to 7 a. m.					7 a. m. to 3 p. m.					3 p. m. to 11 p. m.					Total diurnal heat.	
	120°	105°	90°	75°	60°	120°	105°	90°	75°	60°	120°	105°	90°	75°	60°	C. 2 18	C. 3. 00
50	0	0	1	4	10	49	60	70	77	80	51	40	29	19	11	1149	1580
45	0	0	1	4	9	49	60	71	78	81	51	40	28	18	10	1151	1582
40	0	0	1	3	8	48	61	72	79	82	52	39	28	18	10	1152	1584
35	0	0	0	2	7	48	61	72	81	84	52	39	27	17	9	1142	1570
30	0	0	0	2	6	48	61	73	82	85	52	39	27	16	8	1132	1554
25	0	0	0	2	6	48	62	74	83	87	52	38	26	15	8	1110	1526
20	0	0	0	1	5	48	62	75	84	88	52	38	26	15	7	1087	1492

MARCH 21 AND SEPTEMBER 21.

50 40 30 20	0	0	0	0	3	47	62	76	87	92	52	37	24	12	5	643 766 866 940	884 1052 1190 1292

DECEMBER 21.

50	0	0	0	0	0	45	70	89	97	100	55	30	11	3	0	192	264
45	0	0	0	0	0	46	69	86	96	100	54	31	14	4	0	270	371
40	0	0	0	0	0	47	68	84	95	99	54	32	16	5	0	348	479
35	0	0	0	0	0	47	66	82	94	99	53	34	18	6	1	427	587
30	0	0	0	0	0	47	64	80	92	98	53	36	20	8	2	507	698
25	0	0	0	0	1	47	64	79	91	97	53	36	21	9	2	583	802
20	0	0	0	0	1	47	64	79	90	96	53	36	21	10	3	659	906

As the effect of the sun's heat thus received is to evaporate the moisture at the upper surface of the cloud, rather than to raise its temperature, its tendency is therefore to produce at the upper surface a layer of wholly saturated air whose temperature is little above the cloud mass itself; this latter temperature is given by the law of pseudo-adiabatic cooling that is taking place within the cloud, and is given by the thermodynamic methods of Sir William Thomson (in 1861), and of those who have subsequently developed the subject (see Bezold's memoir in the Berlin Sitzungs Berichte, 1888). But whatever the temperature at the upper surface the increase of buoyancy due to solar heat is simply that due to the buoyancy of the vapor evaporated by it. Now, the latent heat of evaporation of vapor at low temperatures is very nearly 600; in other words, the solar heat expressed in calories divided by 600 gives the number of kilograms of vapor of water that have in this case been formed by the evaporation of the same number of kilograms of the cloud particles. But the kilogram occupies 1,728 units of volume or cubic centimeters, displacing that much air at that level; the buoyancy of one such kilogram of vapor is the difference between its weight and that of the corresponding volume of air or $1728 \times (1-0.622) \times 293$ grams. Pursuing this train of thought we compute the buoyancy to be added to each unit of surface at the top of the cloud, substituting for the exact number some simpler relative numbers, and in plotting them on the map we ascertain whether a special buoyancy over any locality is due to those causes. In general, owing to the location of the sun with reference to the storm center, this additional buoyancy when the sun is at all high in altitude is quite evenly distributed over the illumined part of the cloud surface, and, as the clouds are on the southeast to northeast sides of the storm, with a preponderance towards the northeast, therefore this feature combines with others in causing the storms of the northern hemisphere. On the other hand, as this source of buoyancy has a decided diurnal periodicity, we recognize in it an important reason why storm development, rain-fall, storm winds, and barometric pressure, possibly also storm movement as a whole should have a diurnal periodicity, the minimum of pressure to occur as soon after the maximum of solar influence as is compatible with the inertia of moving masses of air.

(*c*) The third of the preceding additional sources of buoyancy is the heat communicated to the air and moisture in the cloud by the moisture that has fallen as snow, rain, or hail. It will be noted according to the present theory that the mere evolution of heat by ascending and descending vapor only contributes to the ascending current an amount of energy equal to that which is subsequently consumed when the same air descends, as it must descend at some portion of the earth's surface. If there were no loss of heat by radiation and no loss of energy by viscosity and impact, or if the atmosphere and earth constituted a perfect reversible caloric engine or process, then the ascending and descending currents once begun would continue forever; but on account of these losses of energy these currents soon die away ordinarily; and it is when solar radiation contributes its buoyancy to the surface of the cloud instead of to the bottom layer at the surface of the earth that a cloud of any great magnitude is formed, and it is only when, in addition to this, that other free heat is contributed by the fall of vapor from the air that a storm of great magnitude is maintained.

The proper consideration of this latter portion of the heat which goes to make up the storm requires that the daily map should show in much detail and accuracy the distribution of actual rain-fall. But this can not be satisfactorily done without a very much larger number of rain-fall reports than are at present displayed on any daily weather chart that I know of, and the most important addition to be made to the observations depicted on our present charts during the prevalence of a storm would consist in the single item of rain-fall from a large number of intermediate stations, especially in the mountainous regions and slopes of the Appalachian range.

Assuming that we have a sufficient number of rain stations, it is also important that the records should be quite comparable with each other or be reduced to a standard system of accuracy; and this, which has not hitherto been practicable, can be approximately attained by proper methods of exposure of pairs of gauges and reduction of their readings.

Assuming that the weather chart shows a sufficient number of rain-fall reports, we have to recall that these relate to what has happened during the preceding eight, twelve, or twenty-four hour interval; they therefore serve to explain why the present storm center is where it is, rather than to explain whither it will move in the next interval. Nevertheless, we can make use of the reports in connection with the following consideration: The rain falling at A, Fig. 65, will contribute to produce an indraft in the lower air toward the cloud above A, but the indraft and the motion of the distant surface winds require considerable time, except in the case of a vertical ascent, as in a tornado. Therefore, the present position of the low barometer L is connected with the present position of the region of the greatest updraft by a formula that represents a lagging behind, as in the problem of the dog chasing the fox—one is pursuing after the other but never attaining. The path of the cloud may be represented by the curves A_1, A_2, A_3, A_4; the path of the low by the curve L_1, L_2, L_3, L_4, the latter being asymptotic to the former, and if the former describes a very irregular path the latter will cross it frequently, so that the average track of the two will be very nearly the same as in Fig. 66. Thus we see that by plotting on the map carefully the locations of the heavy rain-fall during the past eight hours we obtain the position towards which the air was moving a few hours before and towards which it will continue moving if the location of rain-fall does not change. On the average, American storms move slightly to the right of the location of heaviest rains when the rain is on the

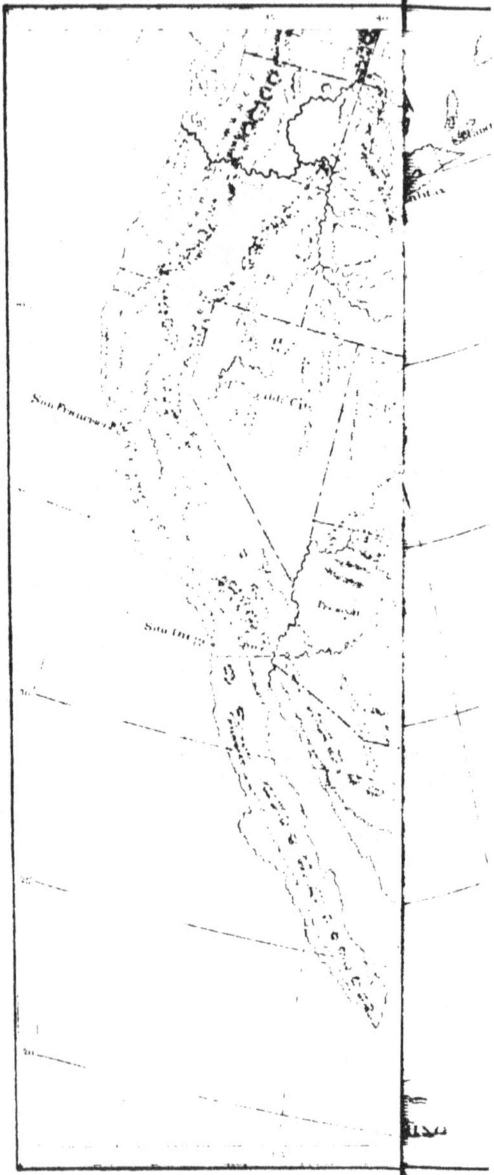

CHAPTER VII.

MECHANISM OF STORMS.

1. We have in the preceding chapters considered the quantity of air flowing horizontally, when the velocity at any point, especially in the cloud, is known, and the bouyancy gradient of flow is known, the latter being the gradient in the direction of the wind, similar to the gradient of flow in streams of water, and not the ordinary meteorological barometric gradient, which is normal to the isobars.

The problem of storm prediction is to determine the amount and behavior of the rising air. We have therefore to consider the condensation or other changes of condition within that air depending on the altitude to which it attains, and this, like the horizontal motion itself, depends again on linear gradients, viz., on the orography, the wind direction, the initial temperature and moisture, and the initial or rate of diminution of temperature or density. The buoyancy developed by the formation of cloud, rain or snow, is now to be increased by the effect of all the solar heat that falls upon the tops of the clouds, and the resulting center of greatest buoyancy is that towards which the lower atmosphere will flow, and about which a whirl, with consequent low barometer, will form.

The prediction of the location of a storm center therefore depends upon the prediction of the location of place of maximum buoyancy; this latter will however be generally moving horizontally at the level of the clouds, while the lower currents, with the resulting isobars, follow on behind it as fast as practicable.

The processes thus generally described are in detail, as follows:

QUANTITY OF ASCENDING AIR.

Assuming that the air moves with uniform velocity for a short time over any region, we must know the orographic and isostathic gradients in the direction of the wind and the resistances that the air experiences due to viscosity and to the convective friction, which latter has two sources, impact and density or buoyancy. The two latter sources vary at the earth's surface too irregularly to allow of being expressed analytically and as they must be considered in any attempt to make quantitative predictions; it is therefore necessary to represent them graphically. I have therefore plotted in the accompanying map of the United States (Fig. 67) my estimated average values of the relative local orographic gradients for winds of each direction, assuming an angle of 1° for the ocean and 2° for the flat low countries; but for winds crossing mountain ranges have assumed an angle of 10° as a maximum for winds of 20 miles per hour; all these angles are doubled for winds of 60 miles per hour, since there is more air thrown up by the obstacle, just as the turbulence of the river depends on its velocity. By the use of such charts and the principles of the preceding chapter it is possible to approximate to the average rate of ascent (*e. g.*, 1 mile in 54 for 1° gradient and 1 in 11 for 5° gradients) throughout a mass of air expressed in terms of the horizontal velocity as a unit. To this orographic effect must be added that due to buoyancy, as estimated from observations of cloud movements or from observation on mountain tops. The velocity of the average mass at cloud level is ordinarily to be deduced from observations of the movements of the lower air by utilizing measurements such as those by Archibald; but when measurements of cloud movements or of balloons are practicable they are of course to be preferred.

Having thus determined the horizontal motion of the air for the average depth up to the tops of the summit, we get the total amount of air that has been drawn into the storm region within a given time, such as the interval between two of the Signal Service maps.

COOLING OF ASCENDING AIR.

Our next problem is to determine the extent of the general process of cooling and condensation that will go on within the air thus transported. The average initial temperature and dew point of such air is given by the weather map, since the mass of moving air over any place has temperature and dew point that depends on the average of those observed during a whole day at that place. The mean condition of the mass changes very slowly by reason of the vertical and horizontal convections; therefore we get a close approximation if we assume that the average temperature and dew point of the whole mass is given by applying to the mean temperature and dew point at the earth a rate of diminution proper for the month. The mean for the past twenty-four hours will serve as the basis for the prediction of the mean for the next twenty-four hours; this subsequent temperature and dew point are to be predicted numerically by table (or graphically by curves) of local diurnal changes at the surface of the earth. The fall in temperature and the precipitation due to the ascent of this air depend slightly on change of latitude and on convection or mixture with surface currents, but principally on the altitude that it attains; the latter effect is given by the thermo-dynamics of the atmosphere as developed by Thomson, Peslin, Reye, and later writers, the general course of reasoning being as follows:

The nearly horizontal movement of the wind carries the lower portions of the atmosphere up over rising land and down the other side.

The observed direction of the wind and of the cloud movement is to be platted for each station, and the computed predicted average velocity and direction for the next twenty-four hours to be laid down. The intersection of these lines with the orographic contour lines will then show the elevations to which the air must have been forced up; hence the ordinary table for cooling, due to ascent, gives the means of computing the average cooling of any part of the whole mass of elevated air, and this may be determined, either from the following numerical table of Hann or by the elegant graphic processes of Hertz (Met. Zeit, 1884). The latter is described in connection with Fig. 58.

The table of Hann and its equivalent in English measures are as follows:

Dynamic cooling of saturated air per 100 meters of ascent.

METRIC SYSTEM.

Initial conditions.

Pressure.	Altitude.	Temperature (centigrade).								
		−10°	−5°	0°	+5°	+10°	+15°	+20°	+25°	+30°
m.m.	*meters.*	°	°	°	°	°	°	°	°	°
760	20	0.76	0.69	0.63	0.60	0.54	0.49	0.45	0.41	0.38
700	680	.74	.68	.62	.59	.53	.48	.44	.40	.37
600	1,910	.71	.65	.58	.55	.49	.44	.40	.37	......
500	3,360	.68	.62	.55	.52	.46	.41	.38	......	......
400	5,150	.63	.57	.50	.47	.42	.38	......	......	......
300	7,430	.57	.51	.44	.42	......	......	......	......	......
200	10,670	.40	.43	.38	......	......	......	......	......	......

ENGLISH SYSTEM.

Pressure.	Altitude.	Temperature (Fahrenheit).								
		22°	32°	42°	52°	62°	72	82°	92°	102°
Inches.	*Feet.*	°	°	°	°	°	°	°	°	°
30	0	0.40	0.36	0.32	0.29	0.26	0.24	0.22	0.20	0.19
25	4,470	.38	.34	.30	.27	.24	.22	.20	.19	.18
20	9,940	.34	.31	.28	.25	.22	.20	.19	.18	......
15	16,990	.31	.28	.25	.22	.20	.18	......	......	......
10	26,930	.27	.24	.21	.19	.17	......	......	......	......

In metric measures.—For air that is not saturated the rate of cooling is 0.9907° C. for absolutely dry air, but 0.9751° for air whose temperature is 30° C. and 60 per cent. relative humidity per 100 meters. (By using Rowland's value of the mechanical equivalent of heat Ferrel gets 0.979° instead of 0.9907°.)

In English measures.—For air that is not saturated the rate of cooling is 0.546° F. for absolutely dry air, but 0.535° for air at 86° F. and 60 per cent. relative humidity for 100 feet of ascent. (With Rowland's value of the mechanical equivalent of heat we get 0.537° instead of 0.546°.)

The cooling thus determined, compared with the dew-point, tells us whether and to what extent the condensation into cloud will take place, but only on the assumption that the cooling mass retains its heat and its mass unchanged as it rises and expands. When the cloud is formed, then to the adiabatic cooling here given must be added the influence of the heat gained by absorption of solar radiation and that lost by radiation from the outer surface of the cloud, as also that lost by the fall of rain, snow, or hail, all of which then become appreciable.

2. The ascending movements in the atmosphere that lead to the expansions, cooling, and precipitation are due to the following ultimate causes:

(*a*) Very local heating of, and evaporation into, the lower stratum and resulting steep vertical currents or interchange of air due to differences of buoyancy, produced by the heat and the moisture, and which differences continue to exist in the ascending mass, relative to its surroundings, until the heat is lost by radiation and the moisture by precipitation.

(*b*) Very wide-spread differences of temperature, such as that between arctic and equatorial regions, plateaus and lowlands, oceans and continents, the dark half and the illuminated half of the earth, these produce a nearly horizontal flow of air underrunning and uplifting the lighter air.

(*c*) The advent of the horizontal flow into a region where the coefficient of horizontal resistance on the earth's surface is increased, such as the flow from the smooth ocean to the land surface or from horizontal smooth prairie to hilly country.

(*d*) The forcible pushing up over hills and plateaus and mountain ridges of air that would have moved horizontally toward a region of low pressure were the ground horizontal. Such cases occur systematically when a region of low pressure advances toward a mountain range. Thus on the accompanying map the low pressure on the west of the Appalachian range is being fed by air which approaches from the southeast, and is being pushed up along the eastern slope of the Appalachians and descending on the western slope.

(*e*) An updraft from the lower stratum is caused when the air immediately above it becomes abnormally buoyant, either by the sudden formation of cloud, rain, and evolution of heat, or by the warming effect of the sun on the cloud. In this case the air below rapidly rises in very steep gradients, not because of its own accession of heat, as in the first paragraph, but because of the increased buoyancy of the air above causing currents in that stratum, and hence a central low pressure up to that altitude, which may be said to apparently draw the lower air up after it, although of course the true motive power is the static pressure prevailing near the ground.

(*f*) Besides the pushing upwards, due to the uniform gradient mentioned in paragraph 4, there is also an important irregular movement when the air passes over hilly countries, due to the fact that the horizontal current impinging against the side of a hill is by its inertia driven upwards; it soon descends again and strikes other hills, and thus any given isobaric or isostathic surface has an undulation similar to the standing waves in a shallow stream flowing over a rocky bed. The interference of these uprising downflowing currents with the ground and with each other causes a loss of horizontal velocity, a thickening of the depth of the horizontal flow, a slight increase of static pressure.

(*g*) The local heatings and evaporation mentioned in paragraph (2*a*) are most active during sunshine and sensibly zero at night-time. These produce in the day-time uprising and conflicting currents and an increase of pressure, such as just mentioned in paragraph (*e*). There is therefore a diurnal period in the force of the wind at the earth's surface caused by this diurnal period in the vertical currents; in so far as the descending air gives up its initial rapid horizontal motion and sets in motion the lower quiet air and overcomes the resistances at the earth's surface there is a corresponding conversion of its kinetic energy into potential energy, and therefore a corresponding rise in the barometric pressure. In the evening as the sun declines and the vertical interchange diminishes there is a less and less loss of kinetic energy, and the barometric pressure at the earth's surface falls to that which is consistent with the rapid motion overhead and the quiet cushion of denser air below. During the night the lower air by radiation cools sufficiently to remain quiet near the earth's surface. It will be observed that the barometric pressure and gradient prevailing in the early morning hours, before the day breeze begins, is therefore the pressure that belongs to the wind velocity prevailing at a considerable distance above, and the gradient during the day-time is one that belongs to a wind velocity very much less than that which is observed to prevail at the earth's surface. This gives rise to a diurnal period in the relation between gradient and velocity that will almost disappear or even be reversed if we use the absolute velocity of cloud movement instead of the lower wind motions.

3. When air is flowing horizontally and begins to be pushed up a plateau or mountain slope by the pressure due to the inertia of that which is behind, the distribution of pressure at various heights on the slope is different from that which prevails vertically at the same height in the original mass above the lowland plains. It can not be said that the particles of rising air lose a portion of their weight by virtue of the ascending motion, since the latter is so feeble compared with the velocity of bodies falling freely under gravity. But, on the other hand, the pressure (p) within the current that is being moved up the slope is equal to that (P) at the same height within the mass of air that fur-

nishes the motor power, minus (1), the effect of converting a portion of this pressure (P) into motion, and plus (2) the effect of again converting a portion of the moving energy into the static energy represented by the obstacles that are to be overcome. In other words, the air ascending the grade (A C, Fig. 68), has to overcome certain obstacles, its own inertia, the viscosity and inertia of the layers under which it is moving and which it is pushing upward and onward, and especially the resistances offered by the ground.

All these obstacles cause a conversion into static pressure or potential energy (p) of the kinetic energy ($\frac{1}{2}\rho v^2$) of the moving mass of air; consequently the pressure at the altitude (h) in the mass A on the mountain slope is greater than the pressure at the same altitude would be were the gas not viscous and the ground perfectly smooth; but the power furnished by the static pressure (P) in the region OO and expended in driving the air up the slope A is $\frac{1}{2}\rho v^2$; therefore the pressure at A is $P-\frac{1}{2}\rho v^2$: therefore so long as motion continues in the nature of a steady push up the slope so long will the pressure at A exceed that at the same height at H by an amount depending upon the relations of the velocity and the resistances at A. If the wind blows steadily for a day for a long time and the pressure at A is observed not to differ sensibly from that at H, we may assume that these terms nearly balance each other. Now, there may be some horizontal motion (V_0) in the mass of air at H, so that the extra horizontal movement at A is the difference between these two and we have

$$P=p+\tfrac{1}{2}\rho(v_0^2-v^2).$$

It can of course happen that the slope may continue up to such a height at C that the pressure at OO is not sufficient to force the air up to it; in this case after the main mass has descended to some such level as M and has pushed the moving mass up to some such level as N, the latter by its greater opposing pressure will then react, and after several oscillations the surfaces of equal pressure will become horizontal and all motion will cease.

This latter result is attained as soon as the resistances to motion due especially to the nature of the surface of the ground have consumed all the initial energy of the oscillating mass. This is what happens at night-time in a closed valley up which or across which the air has been flowing all day; during the day-time by reason of the thermal connection motion was communicated to the very lowest surface layers in the valley, giving them a horizontal movement; with the advent of the evening the vertical and horizontal interchange ceased, leaving the strata to quietly settle down after a few oscillations. One may repeat the experiment with a layer of smoke at the bottom of a bowl of air, (see Fig. 69), and it succeeds all the better if in imitation of the sudden cooling in the evening of the ground we cool the surface of the bowl by setting it in a bath of cold water. The air within the bowl being cool like that in the natural valley can have no motion communicated to it from above no matter how rapidly the upper air may move horizontally except by the intermediation of the viscosity which, as we have before seen, is so slight that the horizontal drag of one layer of air by its neighbor is almost inappreciable.

4. When the air is flowing horizontally over the crest of a ridge its descent on the other side is marked by the following pheuomena: First, the descent is not necessarily very extensive, as one may see by blowing against the side of any obstacle; the mass of air having passed the ridge continues at that elevation, moving nearly horizontally; thus the air may blow, as in our previous experiment, directly across the open top of the bowl without descending into the cold air within; on the other hand if the air does descend and the air below it is displaced to make room for it the descent is by virtue of two distinct phenomena: (1) the elastic force of the upper air drives down the descending air; (2) the force of gravity draws down the denser air. By virtue of the first of these motive powers the static pressure within the descending air is less than that due to its altitude, or rather to its descent, counting from the elevation from which it started to move downwards. By virtue of the second motive power a certain additional amount of kinetic energy is communicated by gravity, and when the air reaches the lowlands this additional component having been transformed into static energy (since the air can not descend any farther) has raised the pressure to its normal value; but as the air descends a sloping mountain side or plateau the conversion is a gradual one, the obstacles to motion are similar to those on the windward or ascending side of a mountain, and, like those, gradually convert the kinetic into potential energy; therefore the pressure at any elevation on the descending side is equal to the pressure at the upper starting level plus the weight of the intermediate stratum minus the pressure consumed in forcing the air downward or in giving additional energy to the downward motion plus the energy communicated by gravity and converted unto static pressure by the resistances to the flow of the air down the mountain side.

In general as the descending air warms up and becomes quite clear the temperature becomes much colder by night near the surface of the ground than was the air on the windward side of the mountain; it is therefore more likely that the lowest layer is in stable equilibrium, and therefore clings to the earth's surface, forming a cushion over which the upper layer slides during the night-time, and exhibits during the day-time less decided vertical interchanges. Thus it happens that the resistances to the descent of the air on the leeward side are less than the resistances to the ascent on the windward side, consequently the air flows faster, and so long as it is flowing steadily down the slope its static pressure is less than at the same level on the windward side; but when it once reaches the base its greater kinetic energy, which is then converted into static pressure, and its greater density combine to make the pressure over the lowlands higher than on the windward side, but over a smaller region than on that side.

The preceding assumes that the air has not lost any of its heat or moisture in its passage over the ridge, and in general for such extended series of ridges as constitute the Appalachian range or the Rocky Mountain plateau region this is approximately true, since whatever heat and moisture have been lost in the early part of the transit are regained in the hours and days required to pass over and begin the descent; but in the narrow ridges that occur among the Swiss Mountains and in individual valleys in our own mountains, where the ascent and descent take place within a few hours, we have to consider the fact that by the condensation of its moisture and its loss as rain while the air is ascending the remaining air has become of an essentially higher temperature than it otherwise would be, so that on its descent it is warmed by compression more than it was cooled when it ascended.

5. Of two masses of dry air having different temperatures, pressures, and volumes it may be that on reducing them to the same pressure and volume they will be found to have the same temperature, thus showing that in their initial condition they had contained the same quantity of caloric; and again of two masses of air that are not dry containing different quantities of vapor, possibly even fog or snow, and having different temperatures, volumes, and pressures, it may be that bringing them both to the same condition as regards pressure and volume they will have the same temperature, since although the quantites of caloric may be different, yet, owing to the differences in the quantities of vapor, these differences will give to the respective masses considered as units different specific heats. If we recognize the fact that the temperature of a mass of air differs according to the conditions we impose upon it, we see the parallelism between such condition and temperature on the one hand and the change in the force of gravity due to the location of the observer on the other. Therefore it has been properly suggested by Helmholtz and Bezold that we speak of the "potential temperature" of a mass of gas, meaning thereby the temperature that it will have under a given standard condition as to pressure and volume. To apply this term to our preceding example we should say that the potential temperature of the air at a given height above sea-level after it has passed the mountain crest and has left much of its moisture there is higher than before it reached the crest. This is the process of reasoning by which Espy (Philosophy of Storms, page XIV, XV, and 241) explained the hot, dry winds that are observed flowing over mountain sides, of which the Swiss foehn and the American chinook seem to be examples.

6. The general question as to the direction in which air will flow in the neighborhood of an obstacle, which is sufficiently difficult when we consider the simplest objects such as a sphere or cylinder in an incompressible fluid extending infinitely in all directions, becomes too diffi-

cult to handle analytically when we consider a compressible gas extending only a short distance upward and bounded below by the rigid surface of the earth. But it is important for us to have as definite an idea as possible of the flow of air around and over mountain ranges, slopes, continents, islands, etc.; there is no difficulty in seeing that when a shallow stream of air impinges against a mountain slope or a ridge that is very long as compared to the breadth of the stream of air, the latter after rising up as far as possible on it will cease to rise and either remain stationary or flow to the right and left. On the other hand, when the depth of the stream of air is very great as compared to the height of the ridge, the latter offers but a feeble barrier, the lower layer of air will be deflected, and possibly entirely retained in circulation on the windward side of the ridge; but the upper greater mass will flow down the leeward side and pass on. Such phenomena are daily repeated before our eyes, especially in the winter time on the weather maps of America and Europe; thus these maps frequently show an area of pressure above 30 inches, representing a deep stratum of cold, dry, and therefore dense air, which easily overflows such ridges as those of the Appalachians, but can not flow as a mass over the higher portions of the Rocky Mountains. This mass of air has usually moved southward for several days attempting to feed a low area on the Pacific coast or in the Gulf of Mexico; it is confined between the Rockies and Appalachians; it can not flow over the former, and therefore it flows south until frequently by uplifting the warmer lighter air it starts a new storm center on its southeastern border and is then drawn off eastward.

7. The clear air and the colder northwest wind in the rear of a cyclone, combined with the warmer southerly winds and cloudy or rainy weather on its front, are the controlling features that decide the rate and direction of its progressive motion. The colder and drier the air is on the rear the more the course turns toward the northeast; the more moisture that is supplied on the east side the more it turns toward the northeast; but the more land and the less moisture on the east side the more it turns toward the southeast.

Storms leaving the North American Continent and advancing eastward over the North Atlantic Ocean have colder northwest winds in the rear at the beginning of their course than when they have reached mid-ocean and beyond. They therefore move more to the north when near the Atlantic coast than when they approach the European coast.

Storms approaching the coast west of the North American Continent have warm moist southerly and easterly winds on their east side, and the mass of moisture that thus feeds them requires more time to attain its maximum condensation and buoyancy than does the relatively small amount of moisture that flows in on the east side when the cyclone reaches the European coast.

The feeding currents over the ocean have a much gentler gradient, since the topographic gradient is 0°, than do the feeders flowing over the European Continent as the storm approaches it.

The result of all the preceding items is uniform and tends to make the locus of maximum cloud buoyancy and solar buoyancy lie farther in advance and farther to the south when the storm is near the European coast than when it leaves the North American coast. This is strongly marked on the Pacific Ocean, where the storms move northeastward along the Japan coast but east or southeast when approaching the Alaska and Oregon coasts. The same result but to a moderate extent is true of the storm tracks on the Atlantic Ocean. Thus Loomis in the revised edition of his Contributions, Part 1, 1885, page 28, gives the result of a number of measures made by himself on the map accompanying the United States Signal Service Bulletin of International Observations; he measured the angle from the north around to the direction pursued by the storm track. These angles as given by him are as follows:

Location.		Angle.	
Latitude.	Longitude.	Loomis.	Corrected.
°	°	°	°
46.9	60	68	66.0
48.9	50	67	65.2
51.3	40	67	65.4
53.9	30	70	68.7
54.9	20	75	74.1
55.5	10	79	78.3

The angles (I^1) as measured by Loomis on the chart need a correction, owing to the peculiarities of that projection, which correction is deduced from the approximate formula:

$$\text{Tang. } I = \text{Tang. } I^1 \cos \varphi^\circ, \frac{57.3^\circ}{90^\circ - \varphi^\circ}$$

Where φ is the latitude expressed in degrees.

The corrected angles are given in the last column, and, being platted on the accompanying map (Fig. 72), they show that on the average between latitude 50° and 55° the storm tracks on the European coast trend about 10° farther south and more nearly eastward than do those on the American coast. As these measures are made upon charts of simultaneous observations, and as the movement is a daily one named at intervals of twenty-four hours, these results are free from any diurnal periodicity in the angle of direction. Such a diurnal periodicity undoubtedly exists and will be revealed by a study that I have begun but not yet completed of the tri-daily simultaneous maps for the United States and Canada.

8. Applying these same principles to storm tracks within the limits of the United States, let us consider the general influence of the Pacific and Atlantic coast lines and the Rocky Mountains and Appalachian slope. Fig. 71 (omitted), map of United States, shows us the Pacific coast (*P.P*), the Rocky Mountain plateau (*RR*), the Appalachian Ridge (*AA*), the Atlantic coast (*TT*), the Mississippi water-shed (*MM*).

A storm center off Vancouver's Island (1) finds its southerly feeder pushed up over the Pacific coast and Rocky Mountains; therefore the progress of the cloud and rain formation is toward the southeast. As the low area spreads over the Rocky Mountain region air flows in from the northeast and southeast, and the lowest pressure is next day found in Dakota or Montana (2) on the eastern slope, but without any well-defined system of rain or whirl; the southeasterly wind forced to rise up the plains of the eastern slope of the Rockies produces its new centers of buoyancy decidedly to the southeast of (2), and an indraft with low pressure becomes central in (3), Indian Territory; the supply of moisture brought by the southerly winds from the Gulf forms clouds that, as they move northward, create new centers of buoyancy more nearly eastward, and the storm center moves to (4) Tennessee. The southeast winds now begin to meet the Appalachian range; so long as (4) is considerably distant to the west of this range the precipitation on the Atlantic coast is not important, and the storm may be said to be fed on its east side by winds descending the west slope of the Appalachians; in such winds no center is easily formed, consequently the northeast winds piling up over the high land of Ohio and Kentucky are more influential, and the center of buoyancy moves northeastward until it reaches the lower Lake regions in (5) Ohio and (6) New York. By this time the southeast winds on the eastern slope have produced sufficient precipitation to start a new center of buoyancy, and the storm moves or jumps over to the southeastward to (7) off the New Jersey coast. In this position the great mass of vapor brought in by both southwest, southeast, and northeast winds produces a greatly increased buoyancy, a more violent whirl, a lower barometric pressure, and, since new condensations may now take place on both sides of the storm center, the latter for a time has a very slow motion forward, or sometimes even back towards (6).

If the north wind is especially laden with moisture there is a strong tendency for the storm to move to the other side of the Appalachian range; such motion has occurred when storms have come in toward New York from the ocean, but not when they have once crossed over the Appalachians from the west. As a rule the precipitation that occurs with east winds blowing along the Atlantic coast upon Newfoundland and Nova Scotia produces such an overwhelming buoyancy in that region that the New England storms move northeastward thither; after these storms have passed beyond Newfoundland the continued precipitation over that island and Labrador deflects them decidedly to the northward. The inertia of the mass of air moving northward over the Atlantic carries it toward Greenland and Iceland, while the buoyancy of the cloud over Labrador and Newfoundland and its continued renewal by the action of cold continental on warm oceanic air checks their eastward progress and even breaks up their existence as cyclonic whirls.

9. Thus we see that in general when a storm center is about to pass over a slope that descends to the eastward the center will move to the southeast, because the southerly winds are ascending and cooling, forming clouds and buoyancy to the southeast, while the northerly winds are descending and drying, hindering the formation of clouds and buoyancy on the north side. The average course of the storms is very much as shown in the sketch (Fig. 71), but for accuracy we refer to the numbers already quoted as given by Loomis at page 33 of the revised edition, chapter I, 1885, of his Contributions. He there gives the average of a large number of measures, showing the angle between the meridian and direction of storm movement at a number of stations in the northwest. These, as platted on the accompanying map (Fig. 72), show that in Manitoba the storm tracks move toward the direction E. 20° S.; when they have reached Kansas or Nebraska, the direction is E. 10° S., but is farther south as the tracks lay farther to the west. For Wisconsin the direction is nearly due east, and so continues over the upper Lakes until it has turned to 10° N. or 20° E. in the St. Lawrence Valley. It then becomes rather southeast until in Nova Scotia and Newfoundland it becomes E. 20° N., and so enters upon the Atlantic Ocean.

In extreme cases the storm centers just passing the ridge of the Rocky Mountains travel rapidly south and southeast to the Gulf of Mexico. This rapid propagation of the low pressure toward the place of greatest indraft and precipitation also occurs on the Atlantic slope; many notable illustrations of it are given at page 34 of Loomis's revised contributions, and are fully explained by the general principles already laid down. This tendency to a southeast movement produces over Kansas and Nebraska, a region where storm tracks intersect each other, and therefore a region of special storm frequency; this fact and its cause was early explained by me, and was especially brought out in the map of lines of equal storm frequency, prepared by me in 1875 for publication in Walker's Statistical Atlas. Similar regions exist over the Lake region and over the North Atlantic Ocean, as shown by Finley's Charts of Storm Tracks.

10. Our study is principally to determine the effects of the moisture, clouds, and the rain. In a dry atmosphere storms do not develop greater severity than shown by the dust whirls of India, Arabia, and Africa, or the horizontal flow of broad sheets of cold, dry air, as in the northerly winds of the Mississippi Valley. In a moist atmosphere the formation of cloud and rain is the principal feature of the development of extended storms. This view of the case is a development of the

views held by Espy, who first demonstrated the importance of the formation of clouds and the attendant development of latent heat; the further prosecution of this idea and the demonstration of the importance of actual rain was a matter of daily practice in my early weather predictions and has been fully dwelt upon by Joseph Henry, Loomis, Ferrel, Meldrum, and more recently by John Eliot and Bezold. The general idea that underlies the present study is perhaps most concisely expressed in the following quotation from the Bulletin of the Philosophical Society of Washington, Vol. 1, page 100, giving an abstract of some remarks by myself in April, 1874:

Professor Abbe stated that the law announced by himself at Cincinnati in 1869 remains abundantly confirmed by daily experience, and might be concisely expressed as follows: A storm center moves toward the region where a given barometric or other condition produces the greatest precipitation of aqueous vapor, or in which the latter is most rapidly taking place. In explanation of this law he added that when we take into consideration all the causes that contribute to produce the precipitation of vapor (whether in the form of haze, fog, cloud, rain, snow, or hail) we are able to account with great accuracy for the direction and velocity of movement of areas of low barometer.

Atmospheric precipitation is produced by cooling the air, and a fall of temperature in any gas is the consequence either of the radiation of heat or of the absorption of heat in performing internal work.

The mechanical absorption of heat is an important feature whenever masses of air are elevated and allowed to expand. This occurs under the following circumstances:

(1) Whenever strong winds blow and in consequence of the inertia of the air and the friction on the earth's surface produce vertical currents;

(2) In consequence of winds being pushed up an inclined plane such as the great plains of the Mississippi basin or the ascent on the east and west sides, respectively, of the Appalachian range and the Sierra Nevada;

(3) In consequence of the elevation of masses of warm air above the masses of cold air, which latter flow, for example, from the extreme northwest southward to the Gulf of Mexico, down the gentle grade of the Missouri and Mississippi Valleys (under this head are included the formation of the local thunder-storms).

The radiation of heat may take place either:

(*a*) Outward into space whenever the air above is dry; or

(*b*) Downward either to the cold earth or to masses of cold air that have underrun the uplifted warmer layers; radiation into space is especially effective after a mass of moist air has been thus uplifted.

The radiation of heat and the visible precipitation of vapor are remarkably counteracted whenever extensive fires prevail by the presence in the atmosphere of very minute particles of carbon or vegetable ashes, which have the property of attracting about themselves quite dense atmospheres of aqueous vapor, precisely as hydrogen is occluded by finely divided platinum.

The direction in which a storm moves being thus dependent to a great degree on the precipitation of moisture, it becomes important to know the location of the sources of atmospheric vapor, especially the presence of regions of snow, forests, swamps, etc., and explanations were given of certain abnormal storm-paths quoted by Professor Loomis.

Especial objection is urged against the idea that high westerly currents carry the storms of America eastward.

11. The general law of upper circulation allows us to form some idea

of the circulation of air going on just beyond the borders of our telegraphic system, but more important is the knowledge possible to be gained fróm the kind and amount of clouds. As illustrating this, I will describe in detail three classes of storms that I have often seen originate and grow in regions where clear weather existed only a short time before:

(*a*) The region of haze and cirrus clouds shows the direction of overflow from distant regions in which a storm is forming or already exists; thus let there be a high barometric pressure on the northeast side of the ridge *AA*, and a low barometer a long distance to the westward or the southward of it as in Fig. 73, which is a sketch of the Appalachian range and Atlantic coast. The figure shows a lower current, *a b c*, flowing from the northeast up the slope, until, reaching the steeper ridge, it is pushed so high up that it finds its path of least resistance to be an overflow down the grade that exists at that elevation in the air forming the slope *c d e f*, which trends to the right of the original lower or surface movement, *a b c*. The cloud observations, unless the reports are too far distant from each other, give a fairly complete history of the changes going on in this moving air; thus from *a* to *b* the air warms a little during the day-time by its southern movement and by the action of the sun and by the mixture of rising, hot, dry currents from the ground, growing drier and clearer, but by 4 p. m. the diminution of solar heat ceases to counteract the cooling by radiation, and at *b* or *c* the air fills with a steadily thickening haze; this increases at night-time over the region *b c*, and especially about sunrise, and reflows as from *c* to *d*, and by early morning mist or rain begins, probably extending in spots along the path or region *b c d e*.

After the sunrise of the next day the insolation warms up the haze and cloud at *c d e*; the indraught along the lower course, *a b*, is increased, and the upper rain bearing stratum extends rapidly from *e* towards *f*, as shown by the reports of haze and rain from the stations to the east of the Appalachian range, but has not yet extended to the very coast itself. The conditions now begin to be favorable for a decided updraught central in the region *d e*, and this perfects itself during the day-time, usually before 2 p. m. of the second day in question. (As an illustration see the map of June 23 to 26, 1889. I recall notable occurrences of this process on November 14, 1871, and November 16, 1873.) The updraught over *d e* is determined when the buoyancy of that region in the upper strata overpowers the movement due to the pressure represented by the density gradients which have thus far been pushing the air from the northeast; as soon as the buoyancy at *d e* produces an indraught greater than that due to these gradients then this buoyancy must be said to be the new and prominent feature in the movement and the former gradients become of less importance. Usually on this portion of the Atlantic coast and generally in all parts of the world, when the local buoyancy thus determines an indraught, it draws the lower air in from all sides, even from across the mountain ridge, and it becomes a new storm center whose existence was owing to the ridge and especially to the winds that were accompanying the former "high" and "low," as depicted on this sketch.

By the velocities and directions of the clouds and winds we shall know that such a new controlling central indraught has been formed. There is first a diminution of wind velocities in certain directions, and then such reversions or such other modifications of these directions as make them conform to a general circulation about the central region. This new arrangement of winds and cloud movements becomes apparent first in the cloud, and is slowly propagated down to the earth's surface until this new circulation is well established among the lower winds, and until then we observe no great fall in the barometer. The uprising of air, or the fall of rain, or the straight-line indraught can, as has been frequently shown, have no appreciable, or rather only a very evanescent barometric influence. The buoyancy can initiate and maintain movements to which the limpid air responds so quickly that a gradient of 0.01 of an inch of pressure for a degree of a great circle will suffice to explain our heaviest winds; but when a movement has once begun in response to such slight gradient there comes into play a centrifugal action and a force depending upon the rotation of the earth, deflecting the currents to the right and leaving a decided depression at the place of central indraught. No sooner has the wind begun to rotate or circulate about this center than there arises still another centrifugal force due to the rotation about this storm center, by virtue of which the barometric pressure is still further reduced, thus giving rise to barometric phenomena as we ordinarily observe them.

In this way our so-called storm center when once started is fed and grows, and if the condensation of vapor, or the sunshine, or the excessive rain-fall, or the descent of dry air on one side of the storm—one or all conspire to tend to develop a buoyancy in the atmosphere on one side of the center greater than on the other side, then the storm will be propagated in that direction. We have here delineated the general history of an important and large class of storms that form on our Atlantic coasts and move northeastward with increasing intensity. The notable subsequent increase in the intensity of these storms when off the Middle Atlantic coast, and the tendency to move north rather than northeast toward Washington and Baltimore, is probably explicable as due to the catch and circulation of a considerable mass of air in the bay indicated by the contours of 1,000 and 3,000 feet on the hypsometric map (Fig. 67), within which Washington and Harrisburg are located. It can even happen that the northwest winds drawn down toward a large storm center shall deposit so much of their moisture or form so large a cloud on the northwest side of the Alleghanies as to cause the storm center to be temporarily transferred northwestward from the Virginia and Jersey coasts even to western New York and Lake Erie and Ontario before it turns again into its northeasterly course.

(*b*) Our second illustration we will take from the weather map of 4.35 p. m., February 6, 1874 (see Fig. 74), showing in the lower Mississippi Valley and westward a region in which hot, clear, dry weather has prevailed for several days. In this case it is the cumulus cloud that first attracts our attention for several preceding days. The 3 p. m. reports show a steadily increasing quantity of cumulus. These die down at night-time, leaving clear skies in the morning, calms at the surface of the earth, and but little general motion in the air overhead; but during the day-time fresh winds prevail and cumuli are formed, showing that an active vertical circulation is going on. As these cumuli increase in quantity and size day by day we argue that the convective process is carrying up to steadily increasing heights the heat and moisture communicated from the soil, and that the layer of warm air is increasing in depth day by day. Finally, on the map of 4.35 p. m., February 5, the report of numerous thunder storms in the afternoon shows that the cumulus masses have ceased to represent a reversible process, and by the rain are leaving in the upper regions an extra quantity of heat, which will probably not all be lost by radiation before it is on the next day re-enforced by an additional portion left by the rain that will fall on that day. In other words, the numerous local vortices represented by isolated cumuli will now combine and give place to a general indraught from all sides of a larger area toward the region where rain has fallen, and a general rotation of the winds will be observed corresponding to a mild cyclone.

If then on this day numerous cumuli again form they will run together into an extended mass of cloud with rain here and there; thus we see on the map for 11 p. m., February 5, a general inflow toward and around the eastern part of Arkansas, and as the mass of air involved in this moves north-eastward we see it forms a well-recognized storm center with central low pressure on the map of 4.35 p. m., February 6.

The original growth of this storm was therefore primarily due to local insolation with the attendant evaporation at the surface of the soil. There was no great movement from any other region into the State of Arkansas, but the air remained there until so broad an area had acquired the necessary buoyancy that it could take upon itself a slow ascent and rotation; when such a region is seen warming up one should watch the growth of the cumulus cloud, and may expect the whole mass to move on within a day or two after numerous local rains have been reported from the cumuli; but when so extensive an area has so decided a buoyancy the barometer falls slightly the fall is in this case apparently a genuine static phenomenon; it need not amount to a tenth of an inch on the average for the whole area, but is indicative of the very important result that follows. This mode of growth of storms in the

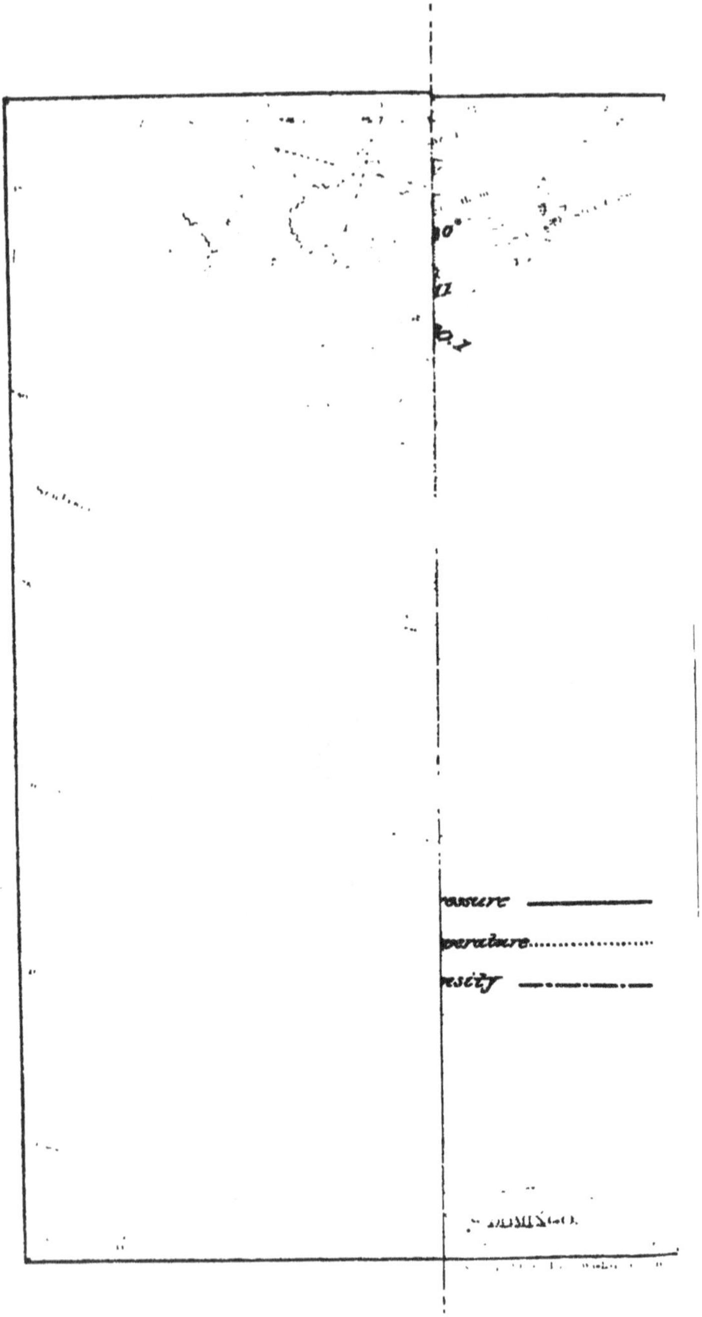

United States is entirely analogous to the typical examples described by John Eliot in studying the origin and growth of the typhoons over the Bay of Bengal.

(*c*) The principle above quoted as to the importance of condensation, that has without exception seemed to me to hold good in all our important storms, must be coupled with the consideration that it is the distribution of density or bouyancy of the air that is largely determined by the observed cloudiness and rainfall; if neither cloud nor precipitation is observed, we shall only be able to study local buoyancy and ascent by determining the temperature at any locality, either at the earth's surface or at a higher level; in the absence of temperature observations at considerable altitudes some use has been made of observed temperatures of the earth's surface. Thus it has been a matter of common observation for many generations that just before a storm, especially a snow-storm, we observe a rise of temperature at the earth's surface. Now, whether this be due to insolation or to the checking of terrestrial radiation by haze and cloud, or to the radiation downwards from warmer air above, it is in either case indicative of increased buoyancy, and should facilitate the ascent of air wherever it is observed, and therefore contribute to the development of a storm.

Therefore, the study of abnormal changes of temperature has from the beginning of my weather predictions been carefully made by me at every telegraphic report in order to detect the first symptoms of the formation of warm layers, and this study has been all the more precisely carried out by the introduction of the departures from normal values. One general rule has been that the temperatures have been found to rise appreciably almost directly in advance of the movement of the storm. In this respect our results differ slightly from those published by Maydell in the St. Petersburg Bulletin of December, 1873, where he concludes that the storm moves in a direction from 40° to 60° to the left of the direction of the greatest rise in temperature during the last twenty-four hours. The reason is partly that we have used eight hour changes instead of twenty-four hour, and partly because our storms are moving from the Rocky Mountain slope or down the Appalachian slope towards moister air instead of toward drier air, as in Maydell's European Studies.

The relation between the density of air and the barometric gradients at low level is shown in the sketch Fig. 75, which represents the storm center of January 13, 1878, 7.35 a. m., and, in which I have drawn lines of equal buoyancy or density or rarity of the air. It will be seen that the denser air is on the northwest side of the center, and that for the heavier air on the colder half of the storm the radial component of the inflow toward the center is nearly equal to the inflow of lighter air on the warmer half. Now, as the latter must rise up in this process, we easily compute how much moisture is elevated into the upper region by summing up the continued product of the average angular velocity in the warmer half by the lengths of the radii of points in that half, by the quantity of moisture per cubic foot; this formula holds good even in a very irregular country and on sloping plateaus or mountain slopes. If we repeat this computation for an upper layer of air representing the clouds, which we may appropriately do by utilizing the observed cloud motions and the map of pressure and temperature at 5,000 feet elevation (see Fig. 76), the average of the figures given for the volume of the inflow of these two layers on the warm half of the cyclone is to be compared with the similar figures computed for the colder half of the cyclone; the latter figures, being the larger, represent the quantity of cold air that is by its underflow displacing and pushing up the warmer air; therefore the difference between the computations for the colder and warmer half represents the moist air whose latent heat, or a fraction of it, is the motive power that is sustaining the cyclone in its growth and that tells us whether, other things being equal, the storm will increase or will die away.

(*d*) In order to determine the direction in which the storm will move it would at first sight seem as though a rigorous law might be attained by substituting in Oberbeck's formula, as deduced in his investigation "On the movements of the atmosphere," a density that (is not constant as assumed by him but that) varies in all directions from the storm center according to some such law as would be shown by moving along the radial lines of flow that cross our lines of equal density, but this will not lead us to the desired result since the process of integration adopted by Oberbeck as well as his original differential formula are based on the assumed constancy of the density and would have a different character if the density had been assumed to vary with latitude and longitude; therefore I will now merely give the general considerations that must be recognized as necessary in order to ascertain approximately on which side of the storm center the greater ascent will take place, and, more specifically, on which side when the ascent has taken place the greater formation of cloud and rain will occur.

(*e*) If the storm center is over the ocean, so that irregularities on the earth's surface are done away with, then assuming that the ascent begins near the radius *C A*, Fig. 77, and that the inflow is at the angle with the isobars shown to obtain for this part of the world, and which may vary with the distance and direction of storm center, we may assume that the rate of ascent above the earth's surface from *C A* to *C D* is sensibly uniform, so that a process of condensation goes on uniformly as the particles rise up in the line on the slope *A. H.*; the *H* is imagined vertically above *D*; the formation of cloud and rain is therefore in this case not dissimilar to that in a slowly ascending and extended mass; the angle of ascent of such slopes is at sea usually less than 2°, and a distance of fifty miles horizontally may be described in order to gain a height of one mile. If the horizontal rate of movement of the clouds has been observed, one can determine

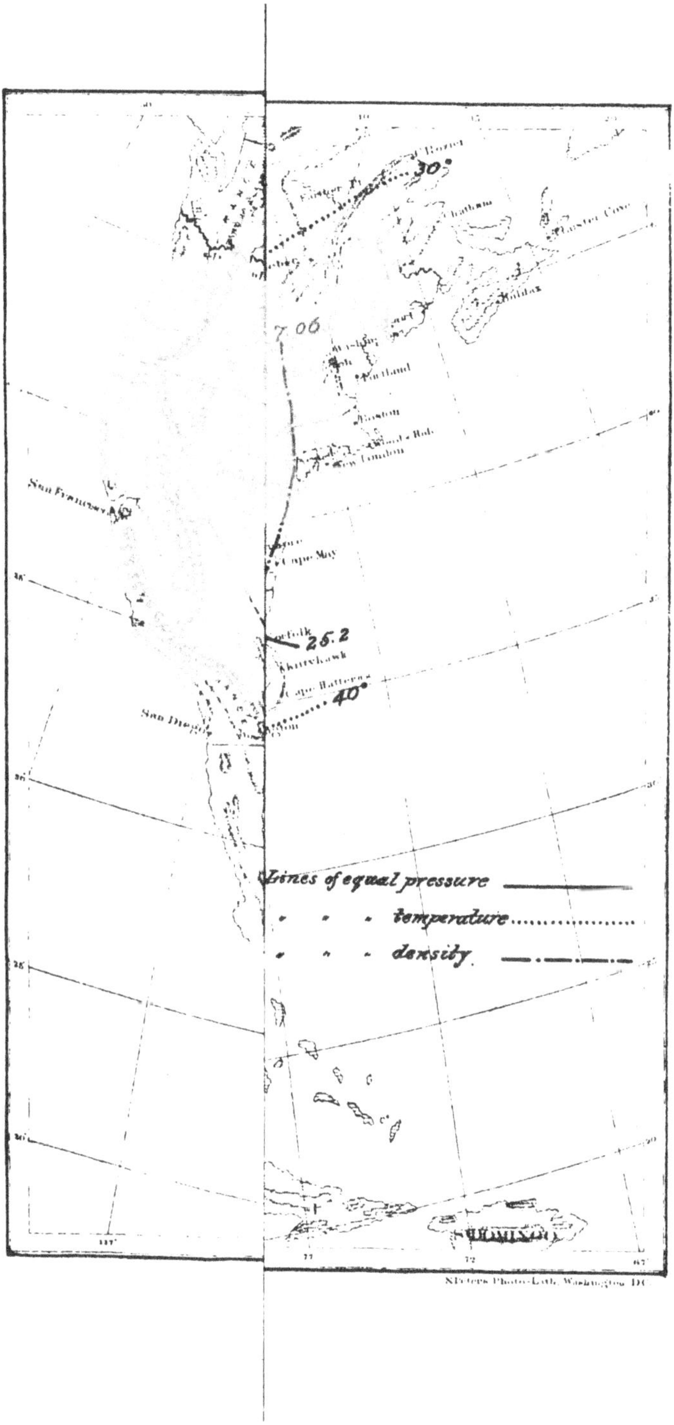
30°
Chatham
Halifax
7.06
Portland
Boston
New London
San Francisco
Cape May
Norfolk
25.2
Cape Hatteras
40°
San Diego
Lines of equal pressure
" " " temperature
" " " density
S. DOMINGO
N.Peters Photo-Lith. Washington. D.C.

how many hours will be required to raise the surface air up to the height necessary for the formation of clouds; the clouds and the buoyancy due to the solar heat falling upon them continue to increase up to 3 or 4 p. m., so that, for example, air starts from *A* at 7 a. m., becomes a cloudy mass of great buoyancy when it arrives at *I* at noon; and air starting from *D* at 7 a. m. becomes a mass of almost as great buoyancy when it reaches *I'* at noon; thus we decide that at noon the place of greatest indraught will be between *I* and *I'*; therefore the storm center will move in a direction between those two points.

Fig. 77.

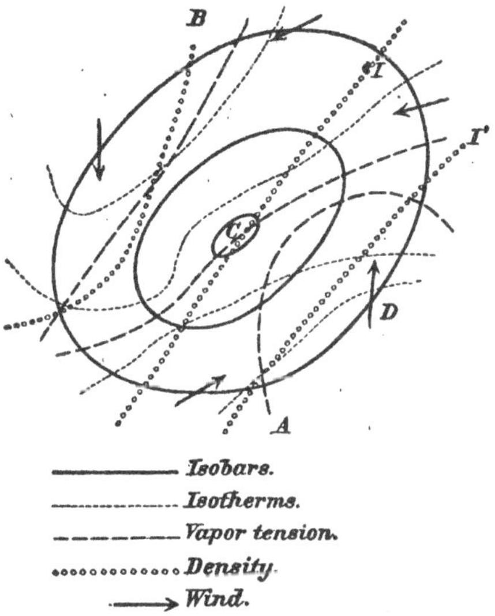

This computation if repeated at a number of points all around the center will show that the air from the north side coming southward has but little tendency to rise, but air on the south and east sides moving northward is certain to rise, and by forming new buoyant centers draws the vortex motion in that direction.

(*f*) Now, the above estimate, comparatively simple over the ocean or level prairie, becomes complicated over the land by the presence of extensive mountain slopes. The difficulty, however, consists principally in allowing for the two special sources of ascent of air: *First*, the heat that reaches the earth and warms the surface layer, which heat varies with the cloudiness and the hour of the day; *second*, the effect of the slope of the land and the ruggedness of its surface, which two features combine to produce the turbulent flow in the stratum of air already discussed in section 2. As regards the effect of topography we shall eventually be able to introduce the co-efficients of vertical buoyancy and of topographic roughness into the formulæ of Hagen, or Boussinesq, or Oberbeck, as has been done for the flow of water in streams; but for the present I will content myself by saying that it suffices roughly to determine the slope up which the air must be forced by the following process: First, make an ideal section (*xy* Fig. 78) through the topography; this gives to the ground surface slope *l m n* in the plane of motion of the wind as it cuts diagonally across the slope of the ridge *r r*. Now, the air close to *l m n* may be considered as a cushion, over which the great mass above will glide. We have therefore simply to add the slope *l n x* or β to the slope due to the buoyancy of the air. These beta slopes, depending upon topography, or, rather, on orography, I have already presented in the chart (Fig. 67) herewith for each of the points of the compass. The betas thus charted represent indeed not merely the geometrical slope, but an additional slope depending upon the width of the ridge and the resistance the air is likely to meet with in crossing over it, and they relate to velocity of 40 miles per hour. The numbers I. have charted are rather arbitrary, and express my own experience rather than any set theory.

(*g*) When air descends on the other side of the ridge the orographic slope "β" becomes negative, and may entirely counteract the buoyant slope "α."

As the sum of the two slopes $\alpha+\beta$ causes a more rapid condensation of moisture, therefore the storm receives an increased tendency to move in the direction of the greatest combined slope, and by taking account of these we shall move the points *I* and *I′*, of Fig. 77, page 149, either northward or southward, as the case may be, and determine the direction where the greatest buoyancy will exist at the end of any given time, hence the direction towards which the vortex will be transplanted.

(*h*) If a cyclonic progress is zero, it must be that the inflowing ascending air affords a sufficient precipitation to maintain the rotation around the center in what is called "steady movement" in Hydrodynamics, that is to say, the buoyancy communicated in successive units of time to the air is exactly sufficient to overcome all resistances to its motion on all sides of the center. If, on the other hand, while the inflow is geometrically symmetrical on all sides, it yet has a chronological periodicity, bringing in sometimes more moisture than at other times, therefore producing a periodicity in precipitation and buoyancy, we shall find the rotary motion and the consequent fall in the barometer to vary periodically and proportionally to the average rain-fall. Occasional examples may be quoted to show the correctness of this view. The most interesting is one quoted by Sprung (page 233, Lehrbuch der Meteorologie), where a cyclone was stationary for seven days in northern Italy, in October, 1877. The barometric depressions and the mean rain-falls near the center of this cyclone day by day were as follows:

Date.	Barometric pressure.		Rain-fall.		Ratio of daily falls.
	Daily fall.	Accumulated fall.	Daily fall.	Accumulated fall.	
1877.	*mm.*	*mm.*	*mm.*	*mm.*	*mm.*
October 4	0	0	1.9	1.9	0
5	2	2	8.2	10.1	4.0
6	6	8	11.8	21.9	1.9
7	6	14	9.2	31.1	1.5
8	7	21	5.0	36.1	0.7
9	15	36	17.8	53.8	1.2
10	7	43	12.2	66.0	1.7
11	6	49	2.3	68.3	0.4
12	....	49	0.2	68.5	....

These figures show that the accumulated rain-fall and the accumulated barometric depressions pursued parallel courses; that is to say, the heat accumulated in the upper air, and which can not rapidly escape by radiation, maintained an accumulating degree of rotation and barometric depression over a large area on the earth's surface; the surface winds needed not to become violent; the upper winds early became strong enough to maintain such a depression without being otherwise apparent at the earth's surface, a phenomenon precisely parallel to what we see in the water-spouts when their incipient vortices depend from the clouds half-way to the earth without at all affecting the surface winds.

(*i*) A third class of storms consists of what has been called troughs of low pressure. A permanent belt of this kind surrounds the earth at the equator, but a perfectly similar belt or trough is very frequently observed in the interior of our continent. Illustrations of such long depressions will be found in May, 1874. These frequently extend northeastward from Arkansas and Missouri to the lower Ohio Valley and the lower Lake regions. Cold northerly winds prevail on the north side and warm southerly on the south side. The trough is not a very elongated ellipse, about which the winds are circulating, but, as shown by the observations of the clouds, the two lower currents, nearly meeting as they do along the line of low pressure, rise up at that point and flow back each on itself. The contrasts between the winds and temperature are exceedingly sharp; thus the charts for 4.35 p. m., May 14, 26, 27, 28, 1874, show a temperature of 50° F. is recorded on the north side, and 80° on the south side, 50 miles distant. In this condition of affairs the northerly winds and their upper return currents from the south deposit little or no rain, and thin haze or clouds prevail for a short distance to the north of the axis of the trough, but the warm southerly winds bring steady rain and extensive clouds for some distance south of the axis. The angle at which the surface winds meet is such that if any rotation were to take place about the trough as a whole it would be contrary to the ordinary rotation of cyclonic storms, but ordinarily no rotation is visible for one, two, or even three days, as in the case of the troughs of 1877, March 22 and June 23–25 (see Fig. 79, omitted).

The whole trough retains its shape and position and oscillates slightly or moves slowly over the surface of the country. It is evident that in such a case the tendency of the orthogonal components of the opposing winds to form a horizontal circulating system opposite to the cyclonic rotation just counterbalances the tendency of the rotation of the earth to force them into a cyclonic system; therefore no horizontal circulation is formed but only a vertical one represented by the two return upper currents (Fig. 80,); consequently the depression at the center of the trough represents the loss of energy taken from the lower wind motion in order to overcome superficial resistances plus the loss due to its efforts to supply the air needed for the upflow due to the buoyancy and updraft at the central cloudy region.

But this condition of equilibrium is quite unstable, and in the course of a few days either northerly or southerly winds or the earth's rotation or the updraft will become the prevailing feature; usually it is the latter that conspires with the winds, and we find a well-marked cyclone forming at some point of the trough, almost always at or near the southwesterly extremity. When and where such cyclone will form is shown first by the cloud movements. So long as the trough exist these movements are as shown in the diagram Fig. 80, in which on the southerly side we find the upper winds moving nearly eastward, but on the northerly side so far as they are visible nearly westward; when this latter motion ceases or the upper clouds moving westward entirely disappear, then we know that the ascending movements on the north side have ceased, that the northerly winds are about to cease rising and flowing back on themselves, but will continue near the ground as their density requires, and will be descending winds rather than ascending, so that they will take the course marked by the dotted line, and will underrun the warm southwesterly winds and establish a horizontal circulation with a cyclonic tendency for the whole system.

12. Besides the preceding special utilization of cloud motions there are atmospheric movements without clouds, or entirely preceding the formation of clouds, that must be attended to in storm studies, especially when a low barometer prevails on the west of the Rocky Mountains or on the west of the Alleghanies and the air on the eastern side just begins to feel its influence and move toward it. Thus, on the accompanying chart for February 16, 1878, 7.35 a. m. (Fig. 81, omitted), we see a flow of cold air over the northwestern Territories drawn

toward the low on the California coast, while at the same time warm southerly winds are blowing over the Gulf States toward that same low. The result is that haze, clouds, and possibly rain form in the southeasterly winds from Arkansas to Nebraska. A new center of indraught is formed there. The northerly winds of Montana deflect somewhat, and, instead of flowing toward the original low, turn toward the new depression, which thus grows and advances over the Mississippi Valley. In this way a storm center is thus transferred from one side of the mountain range to the other. It is very rare that a distinct cyclonic system or a complete storm can be traced in every step of its progress from the Pacific coast to the Mississippi Valley, but much more frequently is it that the original storm whirl breaks up on the west side of the mountains, to be represented by a new storm whirl on the east side. The history of such storms over the Rocky Mountains is repeated on a smaller scale over the Appalachian range, but owing to the moderate elevation of this range it is more common to find, as in the accompanying chart for 7.35 a. m., November 14, 1871 (Fig. 82, omitted), that a storm moving northeastward on the west side of the range gives rise to a subsidiary depression on the east side of the range and some distance to the southeast of the original storm center.

The growth of such subsidiary depressions embraces the following steps: When the original center is at *A* in the Ohio Valley south and east winds blow towards it from the Atlantic coast. These having to surmount the eastern slope of the mountains form cloud and rain on the Atlantic coast, and with the help of the sunshine on the cloud-tops a special indraught is set up toward the region where the cloud and the rain are the heaviest, as at *B*. This indraught is generally sufficiently strong to draw towards itself the air at *C*, giving northeast winds at that point, which would otherwise have had southerly winds blowing toward *AA*. It also draws toward itself the winds at *D* and *E*, or at least prevents that air from flowing toward *AA*, so that in the course of two or three maps or 8-hour intervals the depression at *B* has moved to *BB* and increased to a violent whirl, while the depression at *AA* has dwindled away and finally disappears. This latter is the history of most of the severe storms that develop unexpectedly on the east Atlantic slope, while the observer's attention is directed toward the main center at *AA*. The Appalachian and Rocky Mountain ranges, therefore, act as barriers to divert air from the old storm centers and to cause the formation of new ones. But it is evident that they do not have any material influence of this kind in this direction when the air that is flowing toward the old storm center is also flowing nearly parallel to the ridge, and their maximum effect occurs when the air motion is directly across the ridge and up its slopes.

Within the Appalachian range there are, however, several irregularities, as shown by the contour lines on the hypsometric maps; irregularities such that a given wind, for example the northeast, may have much greater influence at one point than at another in the formation of whirls, clouds, and storm center. If, therefore, on the accompanying orographic chart (Fig. 67) there be represented by figures the relative power at each point of the range that a given wind has, other things being equal, in the formation of a horizontal whirl, it will be at once seen that in the region between Baltimore, Harrisburg, New York, and Philadelphia there is a peculiar tendency to the formation of new centers when the northeast or southeast winds prevail.

13. As the Appalachians are feeble obstacles to the flow of our cold waves, so the Rocky Mountains, and even the plateau of Thibet, are minor matters in comparison with the great ocean of atmosphere. These influence the general circulation of air in the northern hemisphere by introducing into the atmosphere above them, up to altitudes of ten miles or more, vortex and wave motions, standing waves and cloud layers, that then pursue their way onward over the earth, introducing a certain pseudo-periodicity into the formation of highs and lows. Thus we see the familiar barometric areas, with their attending warm or cold, clear or cloudy, descending or ascending currents, circulating with great regularity over the hemisphere covered by the magnificent maps of international simultaneous observations, published by the Signal Service since 1876. The study of these great movements is only possible by means of such charts, and their monthly and annual averages, in which the strictly simultaneous feature, should be uniformly adhered to. These show at once that the secret of the non-periodic variations of seasons and climates in America and Europe-Asia lies in the circumstances of the general circulation, *i. e.*, the connection that must exist between large areas of pressure and temperature prevailing in the atmosphere, and producing motions more rapid but similar to those for the turbulent motion of the river or the ocean.

I have chosen my illustrative maps for storms in the United States from the year 1878, since for that year we are able to make comparisons between the published tri-daily signal-service maps and the published daily international map of the northern hemisphere, the study of which now must deeply engage our attention.

14. The diurnal periodicities hitherto treated of have been shown to produce a diurnal period in the resulting development and progress of storms. The general results as predicted would be as follows:

(*a*) More rapid motion of the storm toward the sun than away from it, namely: southward, if the sun is in the south; eastward during the time that the sun is in the east, and more slowly eastward when the sun is in the west; for temperate latitudes the eastward motion in mid-ocean should be in general greater in the morning and less in the afternoon.

(*b*) More rapid ascensions, heavier rains, heavier winds, larger cloud areas, more buoyancy, more tangential velocity of the wind, and therefore greater central depression of the barometer by day than by night.

In order to ascertain numerically the amount of these diurnal variations, I have made detailed measures of the storms recorded on our tri-daily weather maps, beginning with those of November, 1870, but omitting the first few months, when the stations were not very numerous, I have presented in the following tables the results for the interval March, 1871, to December, 1872, which is as far as my measures have thus far reached. These results, as obtained even in those years, had already established the fact that the movements of our storms are affected by matters that occur far outside of the United States, and that for the proper study of the atmosphere the limits of the weather maps must be extended so as to take in the whole northern hemisphere, an extension which General Myer was able to obtain a few years later.

It will easily be seen that not only the minor but even the major features of storm tracks are due to the general diurnal variations in the synthermal lines, and when the synthermals for the northern hemisphere have been plotted for the Signal Service International Maps we shall see that they are paralleled by corresponding currents in the average movements of storms as to direction and velocity; that, in fact, a cyclonic movement of 1,000 miles in diameter in North America is affected by the temperatures and weather about 5,000 miles away in Europe and Siberia, for a point at latitude 50° N. is only 40° or 50°, namely, 3,000 miles of the great circle distant from the pole and Europe. The study of such large movements within the atmosphere requires the consideration of the movements of the latter as a unit. It is a phenomenon not local but terrestrial, and the hydrodynamic equations must be solved in all their generality, as has lately been approximately done by Oberbeck.

I.—*Diurnal periodicity in the average movements of centers of low areas.*

Month.	Number of—		Monthly averages.							
	Low areas.	Observations.	7 a. m. to 4 p. m.		4 p. m. to 11 p. m		11 p. m. to 7 a. m.		7 a. m. to 7 a. m.	
			Lat.	Long.	Lat.	Long.	Lat.	Long.	Lat.	Long.
1871.			°	°	°	°	°	°	°	°
January	.....	.....	....	....	....	....	....	....	....	....
February	.....	.....	....	....	....	....	....	....	....	....
March	14	33	+0.38	—3.25	+0.50	—2.58	+0.64	—3.29	+1.52	— 9.12
April	10	28	+0.02	—3.16	+0.28	—2.63	+0.61	—2.30	+0.91	— 8.09
May	7	36	+0.46	—2.46	+0.19	—1.25	—0.11	—1.17	+0.54	— 4.88
June	15	43	+0.15	—1.52	+0.07	—1.85	+0.20	—2.25	—0.42	— 5.62
July	16	32	0.00	—2.29	—0.31	—2.19	—0.07	—2.38	—0.38	— 6.86
August	12	33	+0.33	—1.55	—0.14	—1.99	—0.05	—1.94	+0.14	— 5.48
September	15	36	—0.12	—2.07	—0.56	—1.79	—0.45	—1.53	—1.13	— 5.39
October	13	35	+0.50	—3.47	+1.06	—2.44	+0.34	—2.79	+1.90	— 8.70
November	15	38	+0.16	—2.64	+0.33	—2.56	+0.14	—2.66	+0.63	— 7.86
December	15	32	—0.65	—3.50	+0.41	—3.69	+0.80	—4.08	+0.56	—10.67
1872.										
January	12	28	+0.30	—2.60	+0.30	—2.06	+0.52	—2.70	+1.12	— 7.36
February	14	56	—0.31	—3.63	—0.03	—3.26	—0.21	—3.27	—0.55	—10.16
March	19	38	+0.82	—3.04	+0.42	—2.70	+0.85	—3.51	+1.59	— 9.25
April	11	29	—0.12	—3.75	+0.09	—2.86	+0.50	—2.82	—0.47	— 9.43
May	9	33	—0.21	—3.39	+0.50	—2.56	+0.13	—2.50	—0.42	— 8.45
June	9	30	—0.75	—2.48	—0.22	—1.98	—0.36	—1.97	—1.33	— 6.43
July	19	43	—0.41	—2.17	+0.15	—2.44	—0.60	—2.37	—0.56	— 6.91
August	12	44	—0.50	—2.32	+0.18	—1.65	—0.14	—2.26	—0.51	— 6.23
September	14	44	—0.16	—2.59	+0.13	—2.02	+0.13	—2.79	+0.10	— 7.40
October	14	46	—0.34	—1.93	+0.47	—2.09	+0.03	—2.35	+0.16	— 6.37
November	24	40	+0.34	—3.51	+0.24	—2.77	+0.19	—4.12	+0.77	—10.44
December	26	43	+0.45	—4.40	+0.68	—3.83	+0.30	—3.79	+1.43	—12.02

NOTE.—+ is motion *north* in latitude or *west* in longitude.

II.—*Low areas. Solar diurnal perturbations or observed—normal motions.*

Date.	7 a. m. to 4 p. m.		4 p. m. to 11 p. m.		11 p. m. to 7 a. m.	
	Lat.	Long.	Lat.	Long.	Lat.	Long.
1871.	°	°	°	°	°	°
January	......	......	......	......	......	......
February	......	......	......	......	......	......
March	−0.19	+0.17	+0.06	+0.08	+0.13	−0.25
April	− .32	− .13	+ .01	− .27	+ .31	+ .40
May	+ .26	− .63	+ .03	+ .17	− .29	+ .46
June	− .01	+ .59	− .05	− .21	+ .06	− .38
July	+ .14	+ .28	− .20	− .19	+ .06	− .09
August	+ .28	+ .51	− .18	− .39	+ .10	− .11
September	+ .30	− .05	− .23	− .22	− .07	+ .27
October	− .21	− .21	+ .51	+ .10	− .29	+ .11
November	− .08	+ .31	+ .15	− .26	− .07	− .04
December	− .86	+ .50	+ .25	+ .03	+ .61	− .52
1872.						
January	− .12	+ .16	− .03	+ .09	+ .15	− .25
February	− .10	+ .18	+ .13	− .29	− .03	+ .12
March	− .28	+ .43	− .04	.00	+ .32	− .43
April	− .30	− .21	− .05	− .11	+ .34	+ .32
May	− .87	− .22	+ .38	− .09	− .01	+ .32
June	− .25	− .07	+ .17	− .10	+ .08	+ .17
July	− .09	+ .45	+ .40	− .40	− .31	− .05
August	− .31	+ .02	+ .28	+ .17	+ .03	− .18
September	− .20	+ .19	+ .10	+ .14	+ .10	− .32
October	− .40	+ .46	+ .42	− .23	− .02	− .23
November	+ .05	+ .39	+ .02	+ .27	− .07	− .65
December	− .09	+ .11	+ .26	− .32	− .18	+ .22
Sum	−3.15	+3.23	+2.39	−2.03	+0.75	−1.11
Average for one low area	−0.14	+0.15	+0.11	−0.09	+0.03	−0.05

III.—*Average movements of centers of high areas.*

Month.	Number of—		Monthly averages.							
	High areas.	Observations.	7 a. m. to 4 p. m.		4 p. m. to 11 p. m		11 p. m. to 7 a. m.		7 a. m. to 7 a. m.	
			Lat.	Long.	Lat.	Long.	Lat.	Long.	Lat.	Long.
1871.			°	°	°	°	°	°	°	°
January	.....	.....	.....	.....	.....	.....	.....	.....	.....	.....
February	.....	.....	.....	.....	.....	.....	.....	.....	.....	.....
March	12	25	−0.24	−1.33	+0.10	−1.36	+0.30	−2.50	+0.16	−5.19
April	10	28	−0.31	−1.40	−0.38	−1.59	−1.28	−1.40	−1.97	−4.39
May	9	32	−0.88	−1.48	−0.99	−1.42	−1.00	−1.07	−2.87	−3.97
June	17	43	−0.62	−1.03	−0.56	−1.65	−0.38	−1.10	−1.56	−3.78
July	17	52	−0.58	−1.15	−0.08	−0.89	−0.18	−1.18	−0.79	−3.22
August	11	29	−0.12	−1.45	+0.05	−1.28	−0.17	−0.78	−0.24	−3.51
September	13	35	−0.23	−1.82	−0.39	−1.57	−0.49	−2.00	−1.11	−5.39
October	11	36	−0.67	−2.36	−0.10	−1.57	−0.26	−0.86	−1.03	−4.79
November	13	34	−0.37	−2.15	+0.03	−1.06	−0.58	−1.89	−0.92	−5.10
December	11	34	−0.87	−2.33	−0.29	−1.54	−0.45	−1.94	−1.61	−5.81
1872.										
January	10	32	−0.42	−1.73	−0.20	−1.69	+0.03	−1.76	−0.59	−5.18
February	12	36	−0.56	−2.38	+0.12	−1.38	−0.08	−1.77	−0.52	−5.53
March	12	35	−0.85	−1.93	−0.14	−1.90	−0.16	−2.24	−1.15	−6.07
April	9	32	−0.84	−2.74	−0.29	−1.58	−0.23	−1.39	−1.36	−5.71
May	12	45	−0.66	−1.36	−0.22	−2.27	−0.40	−2.22	−1.28	−5.85
June	7	31	−0.58	−1.84	−0.95	−1.91	−0.21	−2.16	−1.74	−5.91
July	13	67	−0.22	−1.13	−0.22	−1.00	−0.18	−0.70	−0.60	−2.83
August	12	65	−0.60	−1.74	−0.37	−1.62	−0.14	−1.99	−1.11	−5.35
September	7	53	−0.22	−2.44	−0.58	−2.45	−0.25	−2.55	−1.05	−7.41
October	13	54	−0.50	−2.48	−0.47	−2.29	−0.16	−3.45	−1.13	−7.22
November	13	31	−0.32	−3.29	−0.67	−3.00	−1.03	−2.61	−2.02	−8.90
December	15	34	−1.25	−3.19	−0.90	−2.65	−0.00	−2.91	−2.15	−8.69

NOTE.—+ is motion *north* in latitude or *west* in longitude.

IV.—*High areas. Solar diurnal perturbations or observed—normal motions.*

Date.	7 a. m. to 4 p.m		4 p.m. to 11 p.m.		11 p.m to 7 a.m	
	Lat.	Long.	Lat.	Long.	Lat.	Long.
1871.	°	°	°	°	°	°
January	......	......	......	......	......	......
February	......	......	......	......	......	......
March	−0.30	+0.62	+0.05	+0.16	+0.25	−0.77
April	+ .43	+ .25	+ .20	− .31	− .62	+ .06
May	+ .20	+ .01	− .15	− .26	− .04	+ .25
June	− .03	+ .39	− .10	− .55	+ .14	+ .16
July	− .23	+ .06	+ .15	+ .05	+ .08	− .11
August	− .03	− .13	+ .12	− .26	− .09	+ .39
September	+ .19	+ .20	− .07	.00	− .12	− .20
October	− .28	− .56	+ .20	− .17	+ .08	+ .74
November	− .02	− .24	+ .30	+ .43	− .27	− .19
December	− .27	− .15	+ .18	− .16	+ .09	.00
1872.						
January	− .20	+ .21	− .03	− .18	+ .23	− .03
February	− .36	− .31	+ .27	+ .23	+ .09	+ .07
March	− .42	+ .35	+ .20	− .13	+ .22	− .22
April	− .33	− .60	+ .11	+ .09	+ .22	+ .51
May	− .18	+ .83	+ .15	− .56	+ .03	− .27
June	+ .07	+ .38	− .44	− .18	+ .37	− .19
July	.00	− .07	− .04	− .17	+ .02	+ .24
August	− .18	+ .27	− .05	− .06	+ .23	− .21
September	− .17	+ .34	− .27	− .29	+ .10	− .08
October	− .08	+ .23	− .14	− .18	+ .22	− .04
November	+ .44	+ .05	− .08	− .40	− .36	+ .36
December	− .44	+ .13	− .27	− .11	+ .72	− .01
Sum	−1.85	+2.25	+0.29	−2.69	+1.59	+0.46
Average for one high area	−0.09	+0.10	+0.01	−0.12	+0.07	+0.02

INDEX.

Page.

Air, ascending currents 27
    buoyancy 102
    dynamic cooling of saturated air per 100 m. ascent 133
    flow of, through tubes 66
    formula for density 102
    formula for flow 8
    formula for mass 103
    horizontal flow 51
    horizontal and turbulent flow 33
    specific gravity 93
    weight of a cubic foot 101
    (*See also* Atmosphere.)
Airy's red and blue magnetisms 17
Allahabad, diminution of vapor tension with altitude 83
    distribution of temperature in lower strata 78
Archibald, diurnal variation in wind 53
    horizontal flow of air 52
Atmosphere, ascending motions 60
    movements 23
    movements of, illustrated by laboratory experiments 29
    section of, around the earth 80
    (*See also* Air.)
Atmospheric electricity, influence on storms 13, 16, 19
Augustin, variations of pressure 38
Auroras, cause 16
Ball, R. S., motion of vortex rings 27
Barometer, origin of diurnal variations 8
Barometric gradient 57
    gradients and density of air 147
    pressure, cause of variations 36
Barus, Carl, earth's magnetism 18
Bassett, circular vortices 28
Bezold, thermo-dynamics of the atmosphere 9
Blanford, variations of pressure 37
Boussinesq, horizontal flow of fluids 8
Bruning, velocities in large streams 50
Buoyancy, cause of ascending motions 60
Buoyancy of air 34, 36, 102
    as a factor in storm prediction 131
    clouds 97, 126, 209
Calms, cause 21
Calories and solar rays 81
    amount of heat in, received from the sun 110, 111
    loss or gain of heat 115

Page.
Cirro-cumulus clouds 86
Cirrus clouds, Espy's view 88
Cloud surface, relative heat received by unit 126
Clouds, average altitude 89
buoyancy 97, 126, 209
cumulus, the summit of ascending currents 71
formation 31, 32, 73
formation of, examples of vortex motions 30
their gradient 85
mackerel sky 86
mare's tails 86
motion 85, 96
Noah's ark 86
obstacles to growth 98
Pele's hair 86
Polar bands 86
"smoking" from the summits of peaks 90
"Table-cloth of Table mountain" 91
velocity 54
Co-efficient of transparency of air 116
Cold waves, origin 77, 125
Cooling of saturated air per 100m. ascent 133
Convective currents 47, 55, 74
equilibrium 71
friction 46, 47
Cumulus clouds, how formed 32
Currents, height of ascending and descending 33
Cyclone, progressive motion 139
Cyclonic circulation, source of its power 77
Declination and inclination of magnetic needle 17
Density of air and barometric gradient 147
formula 35, 84, 102
Discontinuous motion 22
space 22
Droughts and floods, cause 81
Earth's heat, influence on storms 12
internal heat 18
Ekholm, velocity of clouds 54
Electricity, atmospheric, influence on storms 13, 16, 19
Equations of motion for fluids 8, 9
Equilibrium, convective 71
Espy, variations of pressure 37
Evaporation, conditions favorable 117
for different winds 121
Flow of air, horizontal and turbulent 33
liquids, formula 8, 9
Floods and droughts, cause 81
Fluid friction 39
Friction, convective 46, 47
of gases, internal 21
external 22
fluids 39
skin 44
various kinds 61
and vortex motions 20

Page.
Froude, William, skin friction 44
Gravity, correction of mercurial barometers 101
Hagström, velocity of clouds 54
Hagen, horizontal flow of fluids 8
uniform movement of water 48
Hail, how formed 31
Heat, amount absorbed by cloud, haze, etc 112
received at the outer surface of the atmosphere 127
loss or gain, in calories 115
total amount received from the sun 110, 111
Helmholtz, discontinuous motion 22
motion of vortices 23
motions of the atmosphere 9
Helm-wind and helm-bar 91
Hennessey, simultaneous isothermals 78
High areas, average movements of centers 158
solar diurnal perturbations observed 159
Hill, vapor tension at Allahabad 83
Impact, resistance due to 48
Inclination and declination of magnetic needle 17
Internal heat of earth 18
Isostaths 95
Isothermal lines 78
Italy, storm stationary for seven days 150
Kreil, variations of pressure 38
Lamont, variations in atmospheric pressure 37
Latent heat, effect 108
evolution 31
Lightning, cause 16
Low areas, see Storms ....
Lunar heat, influence on storms 12
Mackerel sky 86, 88
Magnetic influences and storms 15, 19
storms, their influence 19
Magnetism, Airy's red and blue 17
declination and inclination of magnetic needle 17
distribution 17
influence on movements of the atmosphere 19
secular variations 16
subpermanent, of the earth 18
Marchand's photantitupimeter 114
Mare's tails 86
Maxwell, molecular vortices 36
Moisture controls the formation, etc., of storms 7
weight of, in a column one meter square 104
Moon and sun, tidal influence of, on storms 13
Motion, discontinuous 22
turbulent 46
Movements of the atmosphere illustrated by bubbles of oil and air in water 23
Noah's ark 86, 88
Oberbeck, motions of the atmosphere 9
Ocean currents, cause 49
Pele's hair 86
Photantitupimeter of Marchand 114
Piezo-electricity 16

Page.
Polar bands 86
Potential temperature of a mass of gas 138
Precipitation, cause of a storm's progress 130
how brought about 31
Pressure variations, cause 36
sources 56
due to vortex motions 39
Radiation, solar, etc., influence on storms 12
Rain, how formed 31
Rain-fall, average value 104
corresponding to a given percentage of moisture 105
stations, need of increased number 129
Rolling friction 21
Rotation of earth, influence 32
Russell, T., evaporation 120
Shooting stars, influence on storms 12
Skin friction 44
Sliding friction 20
Slip, co-efficient 20
"Smoking" from the summits of peaks 90
Snow, how formed 31
Space, discontinuous 22
Solar radiation, influence on storms 13
Specific gravity of air 93
Stars, influence on storms 12
Stewart, Balfour, magnetic fluctuations 19
Storm-centers, diurnal periodicity 9
progress 9
Storms, causes affecting the motion 130
diurnal periodicity of movements of centers of low areas 156
diurnal periodicities 154
direction in which they will move 148
effect of moisture, clouds, and rain 142
formation of subsidiary depressions 153
growth of, depends upon cloud conditions 93
how formed and controlled 7
illustrations of their formation 144
important and unimportant matters 11
influenced by heat of moon, stars, earth, shooting stars, solar radiation, atmospheric electricity, and terrestrial magnetism 12-15
magnetic, their influence 19
mechanism 131
prediction 131
primary sources of energy 11
progressive motion 139
solar diurnal perturbations observed 157
stationary for several days 150
Storm-tracks, changes in types 82
fluctuations 81
in United States influenced by topography 141
rain-fall 143
Sun and moon, tidal influence of, on storms 13
Synthermals 78, 80
variations 155
"Table-cloth of Table Mountain" 91

Page.
Tate, evaporation ... 117
Temperature, distribution of, in atmosphere ... 76
periodic delay in time of maximum ... 100
potential, of a mass of gas ... 138
rate of diminution with height ... 70
resistance due to ... 48
where it is constant ... 75
Thermo-dynamics of the atmosphere ... 9
Thomson, J. J., motion of vortex rings ... 26
Sir William, motion of vortex rings ... 26
turbulent motion ... 46
Thunder-storms, how formed ... 32, 37
Tidal influence of sun and moon on storms ... 13
Tides, solar and lunar ... 38
Tornadoes, how formed ... 32
Transparency of air, co-efficient ... 116
Turbulent motion ... 46
Vapor tension, diminution with altitude at Allahabad ... 83
Velocities in large streams ... 50
atmosphere ... 50
Viscosity of air ... 61
co-efficient ... 20
of fluids ... 21
resistance due to ... 48
Vortex motions ... 20
and cloud formation ... 30
energy converted into heat ... 28
formed in the air ... 41
rings, motion ... 26
Vortices circular with hollow cores ... 28
conclusions regarding motion ... 23
direction of rotation ... 64
Warm waves, how formed ... 125
Water, flow of, through tubes ... 65
Water-spouts, how formed ... 32
Weight of a cubic foot of air ... 101
Wellenmann, formula for evaporation ... 118
Whirls, study of ... 46
Winds, cause of diurnal variation ... 58
effect of obstacles ... 39
upper currents upon lower strata ... 21
variations in force and direction, cause ... 43
velocity at various heights ... 62

o

Spotted something that isn't quite right?

Let us know and we'll fix it for future reprints. And as a thank you for your diligence, we'll send you a free Sothis book of your choice.

publisher@sothis.press

This page intentionally left blank

Except of course for that line above. And this one. And whilst we have your attention, perhaps you'd like to follow us on Instagram @sothis.press. Go on.

Made in the USA
Monee, IL
08 November 2026

68098144R00104